Canada's Weather

Canada's Weather

CHRIS ST. CLAIR

FIREFLY BOOKS

A FIREFLY BOOK

Published by Firefly Books Ltd. 2009
Copyright © 2009 Firefly Books Ltd.
Text copyright © 2009 Chris St. Clair
Image copyright © as listed in Credits on page 225

First printing

Publisher Cataloging-in-Publication Data (U.S.)
St. Clair, Chris
   Canada's weather / Chris St. Clair.
[228] p. : col. photos., ill., maps ; cm.
Includes index.
Summary: Weather from a uniquely Canadian
perspective, from the geographic and topographic
elements that shape weather in Canada, to ways
in which Canadians have turned extreme weather
to their advantage. Also includes basic information on how weather works, weather facts and
Canada's greatest weather events.
ISBN-13: 978-1-55407-338-2
ISBN-10: 1-55407-338-3
1. Meteorology – Canada. 2. Canada – Climate.
I. Title.
551.6571 dc22   QC857.C2S735 2009

**Library and Archives Canada Cataloguing
in Publication**
St. Clair, Chris
   Canada's weather / Chris St. Clair.
Includes index.
ISBN-13: 978-1-55407-338-2
ISBN-10: 1-55407-338-3
   1. Canada—Climate. I. Title.
QC985.S23 2009   551.6571   C2009-901494-7

Published in the United States by
Firefly Books (U.S.) Inc.
P.O. Box 1338, Ellicott Station
Buffalo, New York 14205

Published in Canada by
Firefly Books Ltd.
66 Leek Crescent
Richmond Hill, Ontario L4B 1H1

Cover design by Richard Cote/Sideways Design
Interior design by Gareth Lind/LINDdesign

Printed in China

The publisher gratefully acknowledges the
financial support for our publishing program by
the Government of Canada through the Book
Publishing Industry Development Program.

Contents

Our Obsession With the Weather

by David Phillips

ANADIANS HAVE ALWAYS been interested in and concerned about the weather. Weather has shaped our geography and history, our character and folklore. Weather goes to the heart of our identity as Canadians. It's our birthright, our passion, our national soap opera. Talking about the weather is one of our great national pastimes—well ahead of politics and hockey. Canadians discuss the weather going up or down the elevator, across the backyard fence, at home, at work and on the Internet. Weather is a safe topic of conversation, a way to avoid difficult or controversial subjects. We talk eagerly to friends and strangers alike about black ice and blizzards and in turn listen intently to their stories of frozen pipes and frostbite. "How long did it take you to get home last night?" "How much did your heating bill cost last winter?" "How thick was the ice on your windshield this morning?"

Canadians love and hate the weather (sometimes at the same time!)—and everyone wants to know what it's going to do tomorrow. We have a profound respect for and a passionate interest in meteorology, not because we talk about it more than any other subject but because it's so important to us—always has been and always will be. Part of the appeal of weather is that it is real. No subject compares to weather for its pervasiveness. No other factor, except perhaps our health, looms larger in our daily lives and so directly affects our actions. Weather doesn't merely happen. It happens to our lives and our livelihoods. It touches us, and we touch it. We bless and curse it, chase it and run from it, pray for it and die in it. For city people, weather may be the last piece of nature available to them. On a daily basis, weather influences how we dress, what we eat, how we feel and behave, the cost of heating or cooling our homes and our vacation plans. We have an emotional attachment to weather. It annoys us, disappoints us, terrifies us, threatens us, entertains us and delights us. It presents an opportunity and a challenge, a pleasure and a disappointment. Weather is one of the few things that affect us all. And when something affects everyone, it makes for constant watching and chatting.

It would be difficult to overestimate the importance of weather to our economy. Weather is big business in Canada. About $140 billion of Canada's yearly economy is weather-sensitive. Weather tells us what we can and cannot do. Weather is particularly critical in Canada because of the country's geographical location, its dependence on natural resources and its reliance on export markets. Our polar latitude means that agriculture, forestry and fishing are practised near the northern economic limit. For such sectors, weather can mean the difference between prosperity and bankruptcy. It is often cited as the sole reason for a jump in unemployment, a decline in housing starts, a change in the inflation rate or a rise in the Consumer Price Index. It's almost as if the weather report has become the latest economic indicator. CEOs blame weather for a poor bottom line but are quick to credit shrewd management decisions for positive outcomes.

Canadians are among the most weather-conscious, weather-conversant and weather-sensitive people in the world. We are better educated and more informed about the weather than ever before. I can't think of a field of science other than meteorology that the average Canadian understands better and uses more effec- tively in a whole host of daily activities. Ninety-three percent of Canadians seek out the public weather forecast every day, 365 days a year, and act accordingly. I guess the other 7 percent sniff the air or wet their fingers.

Another reason why weather is so big in Canada is because we're huge, with lots of

weather—maybe too much! Canada is a vast country that stretches across 5½ time zones and covers more than a quarter of the western hemisphere. Vancouver is closer to Mexico City than to Halifax, and St. John's is nearer to Moscow than to Whitehorse. Canada's size and varied geography creates the potential for killer weather. We are a flashpoint between cold arctic air and warm oceanic air. When the two air masses collide, you get weather—often dramatic. In a normal year, Canada sees between two to four tropical storms or their remains; 80 to 100 tornadoes (Canada is the second most tornado-prone country in the world); three million lightning strikes; countless blizzards; face-numbing wind chill; 40°C heat and –50°C cold.

It is said that Canada flourishes because of its diversity. We are also the weather diversity capital of the world. We are much more than a cold, snowy forest. Our climate is a mosaic of most of the climates of the western hemisphere. We have the snow and cold of Siberia; the storms of the United States; the summer heat and humidity of the Caribbean; the aridity of the American southwest's desert; the moistness of Ireland; the winds and fog of Great Britain; and the temperateness of the Mediterranean. Of the world's major climates, only hot deserts and equatorial rainy types are absent.

Perhaps it is the variety of weather that we crave. In many parts of the world,

weather is all reruns. It is so monotonous that it is rarely worth mentioning. How many different ways can you say sunny and fair? Most people don't find their weather very interesting. Who cares about the weather when you can set your watch to it? True, Americans from Montana to Maine are fascinated with weather—but not every American, whereas Canadians' love-hate affair with weather is national in scope and year-round.

Canadians are spared from the world's worst weather—in fact, Canada doesn't hold any world weather records. I continue to be amazed and thankful that so few Canadians die from the ravages of severe weather. It's not that we have a gentle climate. We get our share of weather misery. No part of the nation escapes the whims of the weather gods—the stuff of Hollywood blockbusters. We endure flash floods, weather bombs, hailers, humongous snowstorms, burning infernos and disastrous droughts. It is important that we have a modern weather service and a responsive emergency program across the country, but Canadians are also experienced in weather matters and educated about weather safety. Further, we have a deep and abiding respect for the power of nature.

There are many reasons why Canadians obsess about the weather, but for me, it comes down mostly to its changeability. When Canadians talk about the weather, we speak about how it changes not from place to place, but from time to time. How this afternoon is different than this morning. How this day *this* year is different than the same date *last* year. A comment heard frequently in Newfoundland is, "If you don't like the weather out your front door, look out your back door." Or from the Prairies: "If you don't like the weather, wait five minutes—or drive five miles."

Weather is home to us. Some of our best, deepest and most lasting memories include our experiences with the weather. We often can't remember the weather last month, but bring up the day one gets married or the day of a birth or death of a loved one, and we can recount the weather in surprising detail. Remember the weather the day Paul Henderson scored the winning goal against the Soviet Union in 1972 or the weather that fateful morning of September 11, 2001? Weather can unite us. It is the enemy or the impulse for a random act of kindness. Our weather is something that we all have in common—even if each of us experiences it very differently. The same weather can be one person's misery and another's delight.

For most of the world, weather defines our character and our identity as Canadians. The image of Canada abroad is a land of perpetual winter—a country ice-, snow- and fog-bound at the top of the world. You have heard them. The Great White North, eh! The Land of Snow and Ice. Kipling called us the "Lady of the Snows." Australians call us "frozen Yanks." Our weather gives us a chance to show the rest of the world how tough we are. Canadians like to believe they can survive anything—we brag that there's nothing sissy about living in Canada. Canada is for those in whom the hardy pioneer spirit still burns—people who scoff at blizzards and laugh at frostbite. We have four seasons in Canada, astronomically 91 days long, but climatically much different. There is an old saying that Canada has only two seasons: winter and July 15. But assuredly, we have four seasons, and wherever you go in Canada, it is easy to tell the difference among them.

It's been said that if our climate was different, we would be different. A good case could be made for saying that weather, as much as cultural differences, helps to explain why northerners are different than southerners, and why Prairie people are different than Maritimers. Weather is our most important defining characteristic. The British have the monarchy; the Americans have Hollywood; the French, romance. And Canadians? Well, we've got our weather.

David Phillips is the Senior Climatologist at Environment Canada.

A Land for
All Seasons

THIS BOOK IS about Canada and its weather. I've been fascinated by weather watching for as long as I can remember, and I'm not alone in enjoying this spectator sport. For years as a broadcaster, I've shared my enthusiasm about the weather, explaining how it forms and illustrating why Canadian weather is different from the weather anywhere else on Earth.

So what makes Canadian weather unique? When it comes to climate, as with real estate, it really is all about location.

Canada is the world's second largest country by area—only Russia is bigger. It's not just our immense size, though. Our position on the globe and our topography also combine to create unusually diverse meteorological conditions. Canadians love to complain about the cold—and it can certainly be bitter at the corner of Winnipeg's Portage and Main in January—but we live in a country with so many varied types of weather. Canada is also a land of steamy, humid summers (Windsor, Ontario) and home to a quarter of the world's temperate rainforest (British Columbia). We shovel a lot of snow—folks in Gander, Newfoundland, shovel the most—but we also live in bone-dry cities such as Medicine Hat, Alberta.

As an aviator—I've piloted everything from single-engine aircraft to 737s—I have flown over vast stretches of our country. Even by air,

As the world's second largest country, Canada is host to a wide range of meteorological conditions.

travelling across Canada is an epic journey. Consider that a flight on a commercial airliner from Vancouver to Ottawa takes nearly five hours—jet stream winds at 10,000 metres can reduce this time by up to 45 minutes—and covers over 3,000 kilometres. Unfolding beneath you all the way is an expanse of geography that is simply spectacular, no matter which season you make the flight. From west to

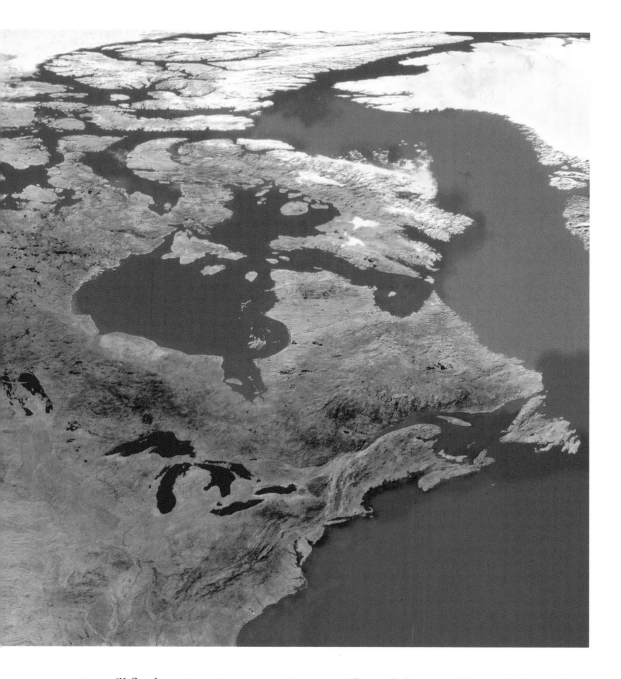

east, you'll fly above towering mountains that are snow-covered year-round, then over a seemingly endless patchwork of plain. Grand old rivers meander through the prairie, dissolving into a boundless forest littered with tens of thousands of lakes. When Lake Superior finally appears on the nose of the aircraft, its size is breathtaking, especially if you arrive at sunrise or sunset, when the light is just right, and the second largest lake on the planet appears as a silver mirror.

Even a trip that covers less than half of this country's breadth—from Toronto to St. John's—is nearly 2,000 kilometres as the crow flies. This route begins on the shores of the Great Lakes, the largest inland lake system on Earth, and takes you over some of the world's oldest mountains. The scenery is spectacular—from rich

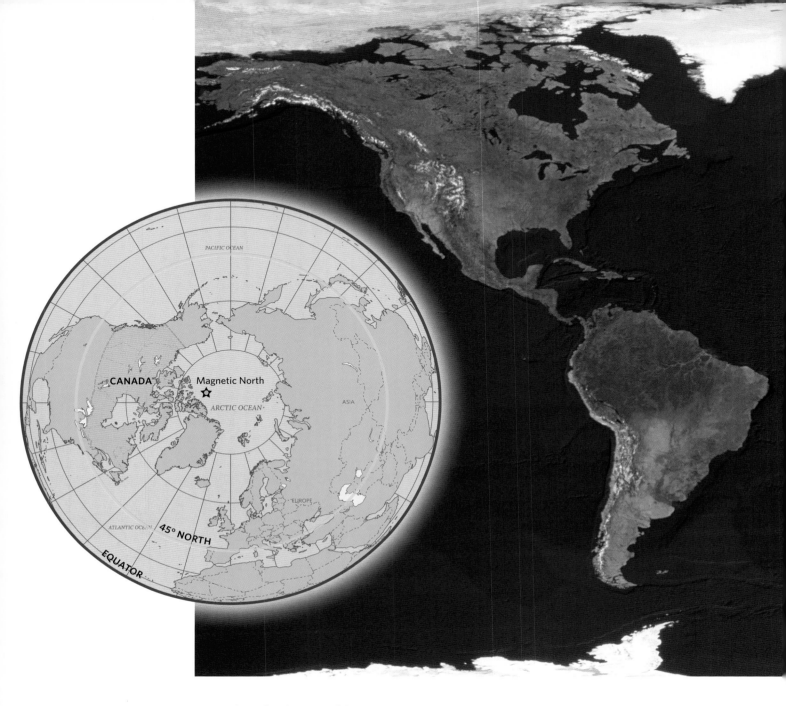

Our Place in the World

As this overhead view shows, almost all of Canada's landmass is located above 45°N latitude. Countries situated near the equator experience far less seasonal variation, since the angle of the sun's rays does not change dramatically over the course of a year.

Countries at mid-latitudes, by contrast, experience much more variation in temperature and hours of daylight. For example, in Quito, Ecuador, the average daily high is between 18°C and 20°C year-round. In Winnipeg, the average high in January is –13°C, compared with 26°C in July.

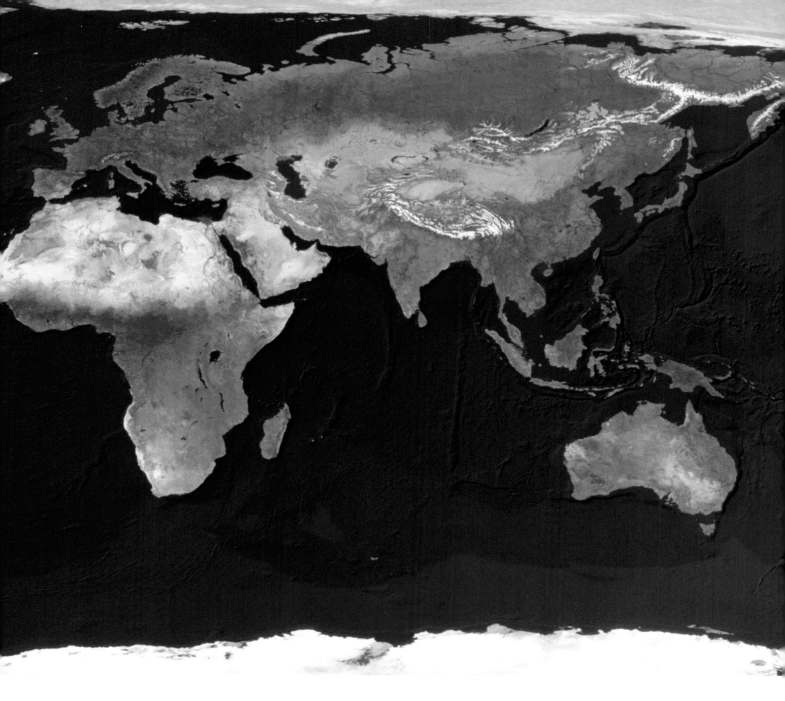

farmland to ancient, rounded hills and carpets of forest that unfold for hundreds of kilometres without any sign of human activity.

These two journeys are extraordinary, and I hope you have a chance to make them at least once. Over the years, I've plotted my many trips on a map, just to see where I have been and what I've missed. One of the gaps in my lifetime itinerary is a part of Canada that many of us never travel over or through: the North. Geographically, what we refer to as Northern Canada is well over half of our country, yet few of us ever get there. It is mostly barren, and it's certainly sparsely inhabited: the population density of Nunavut is 0.01 persons per square kilometre. That's a pretty empty land. But this great open space defines much of our inner spirit. We are a northern people: that's part of what makes us Canadian.

As we'll see in the next chapter, all weather is governed by the heating and cooling of water and land. These are simple laws of nature, but when they're enacted together on a continental scale, they become a fantastically complex and interdependent series of reactions. And in Canada, we have a lot of land and water to heat and cool.

We are located in the northern hemisphere—the half of the planet that is north of the equator. Countries straddling the equator, such as Ecuador and Kenya, don't experience much change with the seasons—the average temperature in December is only a few degrees different from what it is in July. By contrast, the southernmost part of Canada is at about 45°N, or halfway between the equator and the North Pole. People living at these mid-latitudes experience much greater differences between winter and summer. Of course, not all of Canada is in the mid-latitudes: on the contrary, our nation stretches nearly all the way to the top of the globe. The magnetic north pole, in fact, is currently within our borders—all roads may lead to Rome, but all compasses point to Canada.

Our landmass stretches from the Pacific Ocean in the west to the Atlantic Ocean in the east. That's close to 4,800 kilometres from Prince Rupert, British Columbia, to St. John's, Newfoundland. The distance is nearly as great going from south to north: from Point Pelee, Ontario, to Alert, Nunavut—the most northerly inhabited place in the world—the distance is almost 3,900 kilometres.

If you examine an atlas or a globe, you'll find no other nation in either hemisphere that has a physical presence like Canada's—surrounded by three different oceans, stretching from the mid-latitudes to one of the poles and comprising about 10 million square kilometres of land. Let's look at a few other large countries and see how they compare.

The United States covers a slightly smaller area than Canada, but we share a similar topography from west to east, and we both face the Atlantic and Pacific Oceans. The main difference is that the United States has warm oceans to its south: the Gulf of Mexico and the Caribbean Sea. Interestingly, this warm water is a key ingredient of our unique weather as well.

Russia is as far north as Canada, but faces only two major bodies of water: the mostly frozen Arctic Ocean to the north and the Pacific Ocean along its east coast. What Russia is missing is a great body of water on its other flank. Russia has an important Baltic Sea port at St. Petersburg, but most of its western border

Winter storms and cold weather are responsible for about 100 deaths annually in Canada.

Q Where was the coldest temperature in Canada recorded?

A Snag, a small town in Yukon's White River Valley, holds this title. The village was founded in the Klondike Gold Rush of the 1890s. When the Alaska Highway was completed during the Second World War, an airport and weather station were built here. On February 3, 1947, the lowest temperature reading ever recorded in North America occurred there: –63°C.

is jammed against continental Europe: it's more than 2,000 kilometres from Moscow to the Atlantic coast of France or Belgium.

China is much farther south—most of this large nation lies within the same latitudes as the United States—and it faces only one ocean. China's warm seas, which are parts of the North Pacific, contribute to its subtropical climate in the south and temperate conditions in the central regions. Northern China, however, can be bitterly cold and dry, largely because north and west of the country are immense tracts of Asia that virtually cut the region off from oceanic moisture.

Australia is completely surrounded by water—it is the only nation continent on Earth. Australia is in the southern hemisphere, and it's too near the equator to offer the variety of weather conditions that we have in Canada. While its climate varies, it is overwhelmingly an arid continent—only Antarctica gets less precipitation—and does not have meteorologically distinct seasons.

Moving down the list of large countries, most others are smaller than Canada by nearly a third and therefore do not display a similar diversity of weather. It may be more appropriate, then, to compare our country to all of Europe. This continent is as far north as Canada, has almost exactly the same area and has three coasts: on the Atlantic and Arctic Oceans and on the Mediterranean Sea. However, European coastlines are to the north, west and south, whereas our defining bodies of water are to the north, west and east. This is important because of the way global weather patterns emerge—we'll discuss

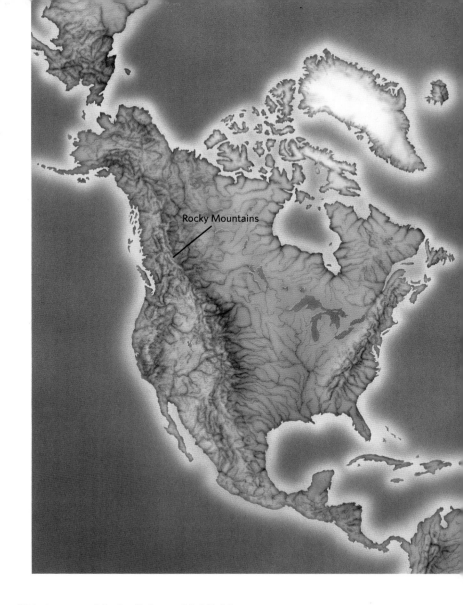

This topographical relief map highlights the Rocky Mountains that the huge size of define our western boundary and extend for hundreds of kilometres eastward.

that topic in our next chapter.

On our western frontier on the Pacific, the coast rises from sea level to more than 5,000 metres. This series of mountain chains, all of which are part of the Western Cordillera, includes the Rocky Mountains, the Cascades, the Columbia Mountains and the Coast Mountains. It is a wall of rock that not only defines our western boundary but also pushes

Above: Besides snow-capped peaks, the Rocky Mountains are also home to a series of picturesque valleys, such as the Okanagan and Fraser.

Opposite: The Great Lakes are thought to be remnants of an ancient freshwater sea called Lake Agassiz, which used to cover most of the Prairies.

eastward for hundreds of kilometres.

These western mountains are relatively young in geological terms. Formed by the collision of the Pacific and North American tectonic plates—two massive slabs of the Earth's crust—they include rock that is almost four billion years old, but most of the peaks we ski on in British Columbia and Alberta are a little more than 65 million years old. Within these mountains lies a series of valleys, including the picturesque Okanagan and Fraser.

On the other side of the Rocky

Mountains, a great plain of flatland rolls eastward. Thousands of years ago, much of this expanse was under water. During the last ice age, which began about 70,000 years ago, glaciers covered most of what we now refer to as the Prairies. When the last of these ice sheets melted some 10,000 years ago, they formed a huge freshwater sea that geologists call Lake Agassiz, after the Swiss-American geologist Louis Agassiz, who first proposed the idea of ice ages in 1837. The Interlake Region of Manitoba and the five Great Lakes are the most conspicuous remnants of this ancient body of water.

Crossing the Canadian Shield to the Atlantic coast of Newfoundland and Labrador, we arrive at another range of mountains, the Appalachians, which are smaller and much older. With peaks reaching nearly 1,000 metres, much of this mountain chain has been eroded by ice ages over the eons. It was formed by the collision of the ancestral plates of Africa against the North American plate some 300 million years ago.

"Don't knock the weather. Nine-tenths of people couldn't start a conversation if it didn't change once in a while."
— Kin Hubbard

Resolute, Nunavut, averages 91 days of blowing snow each year.

Above the 60th parallel lies the forbidding Arctic region. In Canada, the western boundary of the Arctic is the Mackenzie Mountains, an extension of the Rockies. From here, the Arctic runs east over ground that is frozen or semi-frozen for more than half the year. A flight from Yellowknife to Arviat, on the western shore of Hudson Bay, takes you over 1,200 kilometres of scrub forest, marsh, rocky outcrops and hundreds upon hundreds of lakes. During this trip, you'll see little evidence of human activity. It is the hand of nature that has shaped this land with eskers and moraines left by the retreating glaciers nearly 10 millennia ago.

Farther north still is the Arctic archipelago, with massive snow- and ice-covered islands larger than many countries. This region contains mountain ranges as large as some provinces, with peaks reaching up to 2,000 metres. Much of this frozen desert is so dry, so intensely cold and so remote that it feels otherworldly: in fact, there is a site on Devon Island where astronauts train for a possible future mission to Mars.

The southern frontier of Canada is not a natural boundary but a political one—an imaginary line between two countries. Nature does not recognize this border, but the enormous land to the south does play a critical role in our climate.

To understand Canada's weather is to understand its unique position on the planet. We are a northern nation, sprawled across the upper half of a continent, a patchwork of mountains, plains, lakes and frozen tundra, wrapped in the embrace of three oceans. Canada is truly a land for all seasons.

Q **Where does the most rain fall in Canada?**

A **Prince Rupert, in northern British Columbia, receives 2,552 mm of rain every year. The city is also the cloudiest in the country, with 6,123 hours of cloud cover annually (there are only 8,760 hours in a year).**

Bottom right: Above the 60th parallel lies the forbidding Arctic region.

The terrain and glacier features of Devon Island, high in the Arctic, are similar to those on Mars, making it an ideal setting in which to train future astronauts.

How
Weather Works

THINK THAT *why* the weather happens is as interesting as the weather itself. I've been presenting the weather on television since 1994, and there are a lot of ways to handle a forecast. Sure, you can simply read it as a matter of fact, but you can also *explain* it.

I have always tried to explain what's going on in our atmosphere using analogies from our daily life—that's what I love most about what I do. It's how the story of our weather becomes a communal learning experience.

I have something important to acknowledge right way: humans don't understand everything about the natural world. It seems the more we learn about our planet, the more we realize how much is still left to learn. There are natural cycles that we know well, such as the seasons. But there are others that we've only recently figured out, such as El Niño, and there must be countless more that haven't yet revealed themselves.

The good news is that we know a lot more today than we did just 60 years ago. It was after the Second World War that we began to use modern scientific tools to learn about our weather, and during the Cold War, we made great progress in our understanding. Since the 1960s, we've had weather satellites to offer us a glimpse of Earth from orbit, and the technology has advanced to the point where we can now map every square metre of the planet's surface from space. We can also take its temperature and time how fast every patch of Earth heats and cools. The best part is that this amazing techno-logy—which once required supercomput-ers only governments could afford—is easily available to weather forecasters on their laptops.

To understand the basics of the weather around us, you don't need to grasp complex theories of physics—you don't even need a computer. That's not to say that the weather is simple. In fact, the processes that combine to make our weather are extraordinarily complex and unpredictable—that's why forecasts are often inaccurate. Ultimately, though, there are just a couple of important principles to remember. First, weather is simply what's happening in the atmos-phere at a certain location and time. And, second, weather happens because Earth is heated and cooled at different rates.

Of course, the explanation of how and why that occurs is what makes the weather so interesting. In this section, we'll learn about many of the ingredients that play a role in the way the atmosphere behaves. What makes weather so fascinating is that all of these factors are interdependent, and every action creates another action, which in turn creates further actions.

> "Climate is what we expect; weather is what we get."
> — Mark Twain

Q Which Canadian community enjoys the most sun per year?

A The sun shines through mainly clear skies for 2,979 hours each year in Estevan, located in southeastern Saskatchewan.

Our Path Around the Sun

The engine that drives all our weather—not to mention everything else in the solar system—is the sun. At its core, our star has a temperature of about 15,000,000°C—that's an awful lot of energy. As our planet orbits the sun, the atmosphere is bombarded with this energy in the form of heat, light and radiation, and the result is the weather we experience every day.

As we know, Earth takes 365 days to make one trip around the sun. (Actually, it's about 365 days and six hours, which is why we add a day each leap year.) As it travels this path, our exposure to the sun's energy is constantly changing, and because of this, our planet is not heated evenly. That's the first ingredient to creating weather.

The year-long cycle of exposure to sunlight is what we call the seasons, and winter, spring, summer and fall are defined by four important dates: the two equinoxes and the two solstices.

An equinox occurs when the sun appears directly over the equator and provides an equal amount of exposure to both the northern and southern hemispheres. In our part of the world, the autumnal equinox—the first day of fall—occurs on September 21 or 22. The vernal equinox is the first day of spring in the northern hemisphere and happens around March 20. On both equinoxes, day and night are almost equal in duration at mid-latitudes, such as in southern Canada.

The winter solstice, occurring around December 21, is the shortest day of the

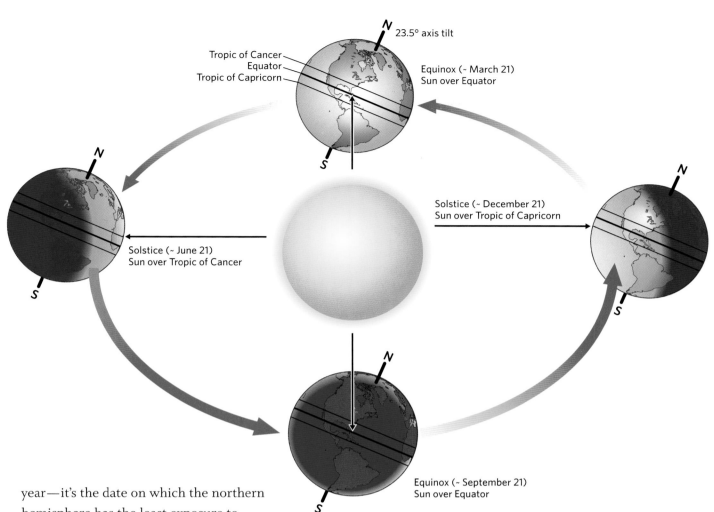

Tropic of Cancer
Equator
Tropic of Capricorn

23.5° axis tilt

Equinox (~ March 21)
Sun over Equator

Solstice (~ December 21)
Sun over Tropic of Capricorn

Solstice (~ June 21)
Sun over Tropic of Cancer

Equinox (~ September 21)
Sun over Equator

year—it's the date on which the northern hemisphere has the least exposure to the sun. The summer solstice falls on or about June 21 and is the day that we get maximum exposure. Even though we may have sunlight late into the evening during July and August, the days have actually been getting shorter for weeks. That's why I love the winter solstice—you really begin to see the daylight hours increasing shortly after it occurs.

At mid-latitudes in Canada, we experience four very distinct seasons with great shifts in temperature. The same city can have summer days that exceed 30°C and winter nights that can be as cold as −30°C. It's all a result of the uneven exposure of our planet to the strongest rays from the sun.

Seasons in the Sun

The tilt of the Earth's axis has an enormous influence on our planet's climate, because it causes the angle of the sun's rays to change over the course of a year. The two equinoxes — during which day and night are almost identical in length — mark the first dates of spring and fall. The solstices — the longest and shortest days of the year — mark the beginning of winter and summer.

Precession

Like a spinning top, Earth has a wobble that causes the axis of rotation to trace out a circle over a period of about 26,000 years. This is called the precession of the equinoxes, and it was first noticed by Hipparchus, the Greek astronomer who lived in the second century BCE. Because of this effect, Polaris's special significance as our pole star is not permanent. This slow movement of the planet may also be responsible for the cycles of ice ages on Earth.

The Earth's Axis

We know that seasons change as Earth makes its annual journey around the sun, but not everyone understands the real cause of the seasons. A lot of people think it's colder in winter because Earth is farther from the sun. That sounds reasonable, but it isn't true. Earth does move closer to and farther from the sun over the course of a year—our planet's orbit is not a perfect circle—but the difference is relatively small. From winter to summer, it varies by about five million kilometres—not much when you consider that the average distance is 150 million kilometres. In any case, Earth is actually *closer* to the sun during the northern hemisphere winter.

So it's not the Earth's changing distance from the sun that causes cold winters and hot summers. Rather, the varying temperatures are caused by the differing angles of the sunlight that reaches Earth. And the reason the angle of sunlight changes is because the Earth's axis—the imaginary line that connects the North Pole to the South Pole—is tipped 23.5 degrees relative to its path around the sun. If this axis weren't tilted but instead were perfectly vertical, the seasons would not occur and the temperature would not fluctuate from month to month.

Every 24 hours—give or take a few minutes—our planet rotates around this axis. At the equator, exposure to the sun is about the same all year long, which is why these regions do not have pronounced seasons. But as you move away from the equator—and because we're talking about Canada, let's move north—the sun's position in the sky changes dramatically.

If you were to go out at noon every day, you'd notice that the sun is much lower in the sky in December than it is in June. That's because from March to September, the northern hemisphere is angled toward the sun, while from September to March, we are angled away from the sun, so we get less exposure to its energy.

This all happens because of that 23.5 degree tilt in our planet.

Why is the axis on a tilt? Scientists believe that when the solar system was formed about 4.5 billion years ago, something—and it must have been something awfully large—bumped our young planet and caused it to tip.

The Earth's axis has another effect on climate, although it's not something you or I will notice in our lifetime. The direction of the Earth's axis slowly changes over time, like the wobbling handle of a spinning top. Today, the northern axis is pointing almost directly at the star Polaris, but in about 13,000 years, it will point to the star Vega. Then, roughly 13,000 years after that, the axis will come full circle, and Polaris will be back in its position as the pole star. Many climatologists believe that this cycle, which is called the precession of the equinoxes, creates ice ages on Earth.

Q When was the first weather forecast issued in Canada?

A On September 4, 1876, the first storm warning prepared at the Toronto Observatory was issued to the public. In October of that year, the first regular general forecasts began.

Land and Water

You might be surprised to learn that around 70 percent of our planet is covered by lakes, rivers and oceans. Almost three quarters. This ratio of water to land is a primary building block of our weather.

Have you ever noticed that when you spend a hot summer day at the beach, the sand gets amazingly hot by mid-afternoon, yet the water somehow seems cooler? Then when evening arrives, the sand feels cool, but the water seems warm. What causes that?

The explanation is that it takes fewer calories (units of energy) to heat land than it does to heat water. Introducing calories of energy gets molecules to vibrate and to emit what we feel as heat. In a solid—such as the sand on the beach—the molecules are more densely packed than in water, so the energy transfer from molecule to molecule is much easier and quicker. In a liquid, the molecules are farther apart, so energy transference is a much slower process. That's why beach sand heats up faster than the water.

When the heat source is removed—when the sun sets at the end of the afternoon—the molecules that make up the land rapidly lose energy. The action of the molecules bumping against each other expends energy and causes the motion to slow down, so less heat is emitted. Meanwhile, our water molecules, which have been slowly heating up all day, also begin to lose the energy the sun was providing. But because these molecules are farther apart, they take a longer time to slow down their motion. So the water stays warmer longer than the beach sand.

During the vernal equinox, the sun is directly above the equator, and day and night are equal in length.

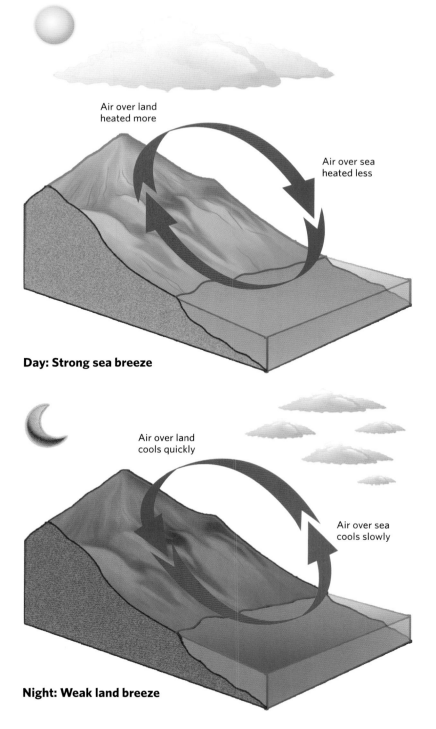

Day: Strong sea breeze

Air over land heated more

Air over sea heated less

Air over land cools quickly

Air over sea cools slowly

Night: Weak land breeze

Winds of Change

The uneven heating of the Earth's surface is one of the engines that drives weather systems. One example occurs in coastal areas, where land and water absorb and lose heat at different rates. During the day, the air over the land heats up more quickly than air over the sea, and the pressure difference generates strong sea breezes. At night, the air over the land cools more quickly, and the wind pattern reverses.

This complex process is repeated all the time, all over the planet. But keep in mind that there's far more water than land on Earth, and all that water serves as a great reservoir of heat.

The Lay of the Land

Many ancient cultures believed the world was a flat disk in space. By the time of the first-century astronomer Ptolemy, however, most educated people accepted the idea of a spherical Earth. The fact that Earth is an orb is integral to the weather, since a round planet cannot be heated evenly. The land nearer the equator receives much more sun than the land at higher latitudes.

So Earth isn't flat or smooth: it's an uneven quilt of textures, from rocky, ice-covered mountains and flat expanses of sand to carpets of rainforest. The lay of the land—and what's covering it—is very important to how weather patterns emerge. The composition of the Earth's surface affects how it absorbs, retains and reflects the sun's energy.

When you wear dark clothes in the summer, such as a pair of black pants, you feel hotter than when you wear something light-coloured. The reason is that lighter colours reflect the heat of the sun. Darker colours, however, absorb and retain heat more readily.

With this in mind, look at an image of Earth from space. Parts of the land are covered in white ice, while other regions are blanketed with dark green forests. Ice reflects much of the sun's energy, while a dense forest absorbs and holds on to much of that heat. On a much smaller

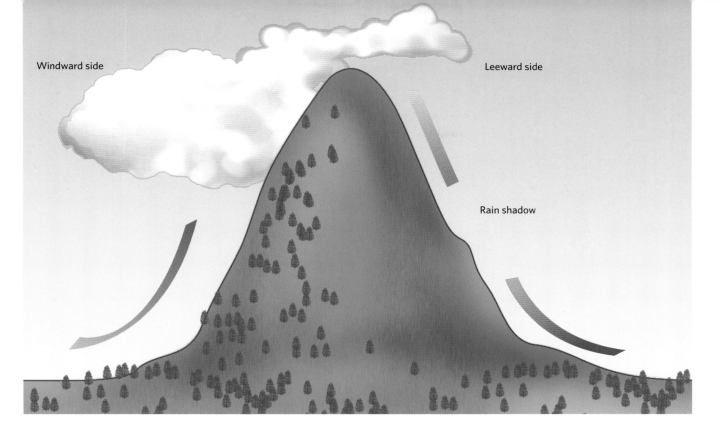

Windward side

Leeward side

Rain shadow

scale, urban areas become warmer than the surrounding countryside because there is more asphalt and more dark roofs to retain heat.

The topography of the land also directs what happens in the atmosphere. As wind moves across the surface of Earth, mountains and valleys will alter its direction. Mountain ranges force moving air upward; this is called upsloping, and it plays an important part in Alberta's stormy weather. After the wind reaches the top of the mountain, it loses moisture, and a warm, dry wind rushes down the leeward slope—in western Canada, this is known as a chinook. This flow of air over mountains can lead to great changes in temperature and pressure.

Valleys cause winds to swirl and increase in speed. They also heat air masses much more efficiently than an open plain—the air in a valley is warmed from above by the sun and by the heat radiated from the floor

Rain-Shadow Effect

When warm, moist air is forced against a mountain or large hill, it rises, cools and condenses into clouds. This process often results in heavy precipitation on the downward side of the slope. The air that descends the leeward side, however, is warmer and dryer, creating a "rain shadow" where precipitation is far less frequent. The semi-arid climate of British Columbia's Okanagan Valley, which is on the leeward side of the Coast and Cascade Mountains, is a result of the rain-shadow effect.

and two walls, similar to the way an oven directs heat toward a roast. That's why those rugged valleys in British Columbia often see temperatures over 40°C in the summer. The terrain is, in effect, "cooking" the atmosphere.

Plains, on the other hand, offer an ideal platform to generate large areas of both stable and unstable air above the surface. In summer, the rapid heating and cooling can create massive thunderstorms. In winter, the great reflective surface presents an ideal opportunity for strong high-pressure cells and a further intensifying of the cold.

Precipitation

Condensation

Evaporation

River discharge

Surface run off

Groundwater flow

The Water Cycle

The Earth's water is continuously redistributed through a series of processes called the hydrological cycle, or, more commonly, the water cycle. Water evaporating from oceans and lakes rises and condenses into clouds, which return the water to the surface in the form of precipitation. Some of this rain and snow is absorbed into the ground, while much of the rest makes its way back to the sea via surface runoff and river discharge. Plants also play a role by absorbing water and returning it to the atmosphere through transpiration.

The Many Faces of Water

Since land makes up only about 30 percent of our planet's surface, that leaves 70 percent which is covered by lakes, seas and oceans. This enormous amount of water is an active ingredient in our weather. All this water is being heated or cooled at different rates and, as we've seen, uneven heating has a number of effects on the atmosphere.

Because Earth is a sphere, the water nearest the equator should be warmest, and the water closer to the poles should be cooler. In a broad sense that is the

case, but things aren't that simple—not all water is created equal.

Look again at our image of Earth from space. All the blue is water. So is all of the white. And so are those swirling clouds that surround the planet. Indeed, one of water's most intriguing characteristics is that it's the only substance which is commonly found in three forms: liquid,

Water is found in liquid, solid and gas forms, and all three of these states are visible here.

Almost three-quarters of the surface of Earth is covered by water, but its depth, salinity and exposure to the sun differs greatly, which results in different effects on weather.

solid (ice) and gas (water vapour).

About 70 percent of the world's fresh water is frozen in ice caps and glaciers. This solid water plays a role in weather by reflecting much of the sun's energy back into the atmosphere. It also acts as the Earth's refrigerator, cooling the air and water that surround it.

The clouds that grace our skies are also water. There are literally billions of tonnes of droplets over our heads. The amount of this moisture in the atmosphere has a profound effect on local weather.

Of course, the vast majority of our water is in liquid form. Here again, though, not all liquid water affects the weather in the same way. To begin with, the salinity of the water—how much salt it contains—is crucial. Of all the water on Earth, 97 to 99 percent is seawater. Salt water freezes at a lower temperature than fresh water, and it takes a higher temperature to evaporate it.

The depth of the water is also an enormous influence. Shallow water heats faster than deep water, which is why Lake Winnipeg is warmer than Lake Superior. You can do this experiment at home. Pour some water onto a cookie sheet, then pour the same amount of water into a mixing bowl and put both outside in the sun. The water on the cookie sheet gets warmer much faster than the water in the bowl.

What all of this means is, because they have various depths, levels of salinity and exposure to the sun, large bodies of water have different temperatures and different rates of evaporation. These variations govern much of what is happening in the atmosphere.

There are four layers in Earth's atmosphere: the thermosphere, mesosphere, stratosphere and troposphere. All of our weather occurs in the troposphere, the lowest layer.

The Atmosphere

Our atmosphere is about 21 percent oxygen and 78 percent nitrogen. The remainder is tiny amounts of other components: argon, ozone, water vapour and carbon dioxide, to name a few. Compared with the radius of our planet, the atmosphere is extremely thin. In its entirety, it extends less than 600 kilometres aloft—about the distance from Montreal to Toronto—but about 99 percent of its mass is contained in just the first 30 kilometres.

The atmosphere has four layers. The **troposphere** begins at the surface and extends about 15 kilometres high at the equator (it's thinner over the poles). The air is most dense here—in fact, 75 to 80 percent of the atmosphere's mass is contained in this layer. The temperature drops steadily as you move up—at the top, it's colder than –50°C. The troposphere is where we live, fly airplanes and experience weather.

The warming and cooling of air in the troposphere governs almost all of meteorology, and it's all based on a simple principle: hot air rises and cool air sinks. Because the molecules are farther apart and moving more rapidly, warm air is less dense and therefore lighter. In cooler air, the molecules become compacted, as we discussed earlier. When this happens, the air becomes heavier, and it sinks toward the surface.

The cycle of heating and cooling in the atmosphere begins simply: as the Earth's surface is heated by the sun, the air above it is also heated. That warm air rises, and the void it leaves is filled by cooler air that circulates down from the atmosphere. Now the new, cooler air is heated, and the cycle is repeated—all day, all around the planet, over land and sea.

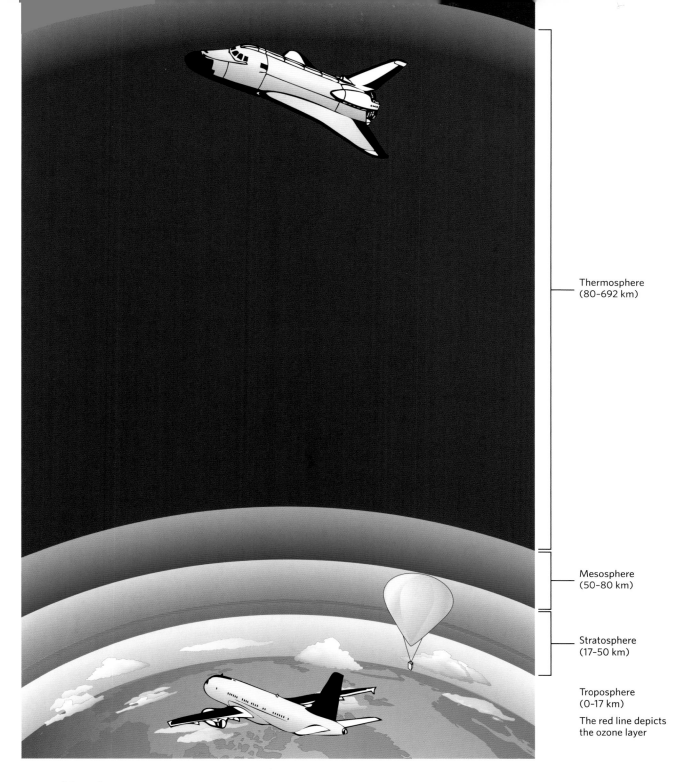

Thermosphere
(80–692 km)

Mesosphere
(50–80 km)

Stratosphere
(17–50 km)

Troposphere
(0–17 km)

The red line depicts
the ozone layer

Into Thin Air

The atmosphere wraps our planet in a blanket of oxygen, nitrogen and small amounts of other gases. While the atmosphere is hundreds of kilometres thick, more than three-quarters of its mass is contained within the troposphere, which reaches no more than 15 km above the surface. This is where our weather happens. The temperature of the troposphere declines rapidly as you ascend, from an average of 15°C at the surface to below –50°C at the tropopause, the troposphere's upper limit.

A low-pressure system is responsible for the impending precipitation at this isolated farm in Elmira, Ontario.

to spiral counterclockwise. In a high-pressure system, however, the stable air that settles toward Earth slowly rotates in a clockwise, or **anticyclonic**, direction. (By the way, the Coriolis effect does not cause water to spiral down the drain in opposite directions in the northern and southern hemispheres. That's a well-travelled myth.)

Low pressure is also associated with precipitation. As the air in a low-pressure system ascends, the moisture condenses. When these water droplets are very tiny, they form clouds, but if there is enough moisture, the droplets will grow large and will fall back to Earth as rain—or, if the temperature is low enough, as sleet, hail or snow.

Air Currents

Just as ocean currents circulate water around our planet, air currents keep our atmosphere in constant motion. These air currents, like so many other weather phenomena, are driven by the uneven heating of the Earth's surface.

To understand how air circulates in our part of the world, imagine the northern hemisphere divided into three horizontal slices, each consisting of 30 degrees of latitude. The air above the southernmost section organizes into patterns called Hadley cells. Warm, moist air above the equator rises to the troposphere, then it moves north, cooling as it does so. Much of this air then settles back toward the surface at about 30°N. The air near the surface, steered by the Coriolis effect, then moves toward the equator, where it is warmed again, creating a continuous loop. The northeasterly air currents created by Hadley cells are called the **trade winds**.

In the top slice of our hemisphere—from the North Pole down to 60°N—air circulates in what are called polar cells. These aren't as strong as Hadley cells, but they behave in a similar way: cold air near the pole descends and moves southward until about the 60th parallel, where it warms, rises and moves back toward the pole to continue the cycle. All the while, the Coriolis effect swirls these air currents into prevailing winds called the **polar easterlies**.

Between 30°N and 60°N—which includes just about all of the inhabited regions of Canada—things are more complicated. Atmospheric circulation here occurs in what is called a Ferrel cell. Unlike in Hadley and polar cells, the air doesn't circulate in continuous loops. As a result, while the air near the surface tends to move from southwest to northeast—these are the prevailing winds we call the **westerlies**—the system isn't very stable. Wind direction and intensity can change quickly and dramatically. This causes a mixing of cold, dry air with warm, moist air, and that's one of the reasons why the mid-latitudes have more variable weather than regions near the equator or the poles.

Near the boundaries between high- and low-pressure areas, we find the jet streams. Jet streams develop about 10,000 metres above the surface, at the top of the troposphere. The air here is warmer than the air in the stratosphere above, and the difference in temperature creates a difference in pressure. We've already seen that winds are caused

A star's twinkle is caused by shifting air currents in the atmosphere.

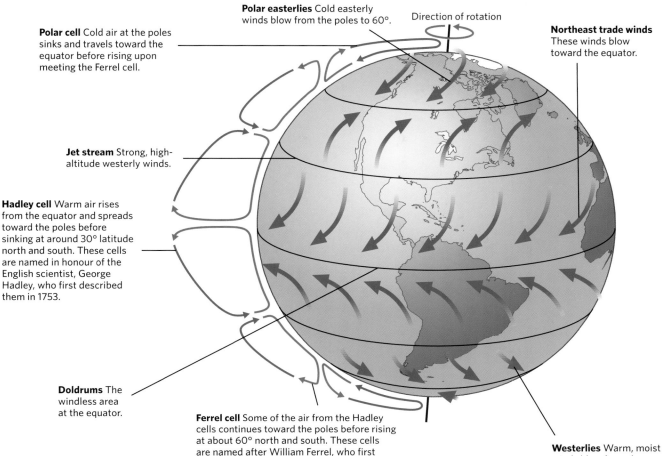

Polar cell Cold air at the poles sinks and travels toward the equator before rising upon meeting the Ferrel cell.

Polar easterlies Cold easterly winds blow from the poles to 60°.

Direction of rotation

Northeast trade winds These winds blow toward the equator.

Jet stream Strong, high-altitude westerly winds.

Hadley cell Warm air rises from the equator and spreads toward the poles before sinking at around 30° latitude north and south. These cells are named in honour of the English scientist, George Hadley, who first described them in 1753.

Doldrums The windless area at the equator.

Ferrel cell Some of the air from the Hadley cells continues toward the poles before rising at about 60° north and south. These cells are named after William Ferrel, who first identified them in 1856.

Westerlies Warm, moist winds blow from the west.

when a stable, high-pressure air mass butts up against a volatile, low-pressure one, and the pressure differences are greatest right at the boundary of these two atmospheric layers. The jet streams meander their way along this boundary, usually moving to the east because of the Coriolis effect.

A jet stream may be thousands of kilometres in length and over a hundred kilometres wide, but it is only about a kilometre deep. The best way to envision it is to imagine the shape of a flattened cylinder, in which the velocity of the wind increases as you draw nearer to its core.

Each of the jet streams is more prominent during certain times of year. In Canada, we're most affected by the polar jet stream, which creeps south of the 60th parallel in winter. When strong areas of low pressure interrupt the west-to-east flow of the polar jet stream, it can diverge, creating a second current called the arctic jet stream. The polar and arctic jet streams define the southern limits of cold air from the north. In the summer, we may also be affected by the subtropical jet stream, which moves north from the 30th parallel, marking the northern boundary of the warm and humid air from the tropics.

In the warmer months, jet streams can interact dramatically with powerful thunderstorms. When a storm grows very rapidly, the volume of rising warm

The Trade Winds

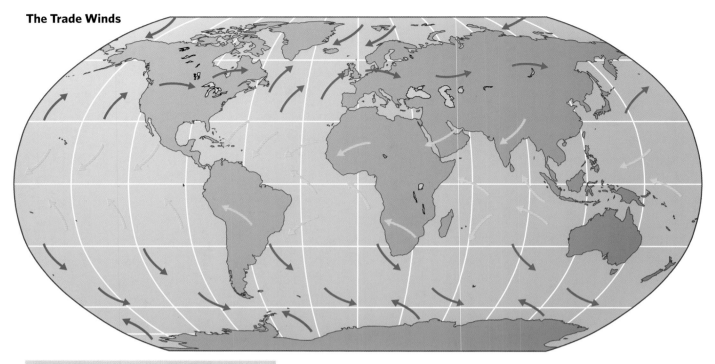

Global wind patterns
←— Polar easterlies
—→ Westerlies
Northeast trades
and southeast trades

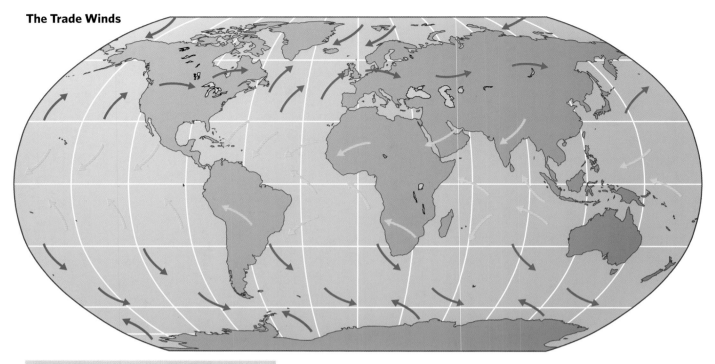

Fog is defined as air droplets in the atmosphere that cause visibility of less than 1,000 m.

air can be so great that it actually draws cold air from the jet stream toward the surface to replace it. This is a large part of the dynamic that creates micro-bursts—columns of descending air that produce straight-line winds as opposed to swirling ones—and even tornadoes.

It's often said that low-pressure weather systems are steered by jet streams, but it's more accurate to say that they're perpetuated by them. Jet streams provide the temperature contrasts that these weather systems need to exist. When

---

Q Which Canadian city is windiest?

A St. John's, Newfoundland and Labrador, is the windiest city in Canada, with an average annual wind speed of 24 km/h.

---

those temperature contrasts decrease, so does the strength of the jet stream and the intensity of the low-pressure system.

Ocean Currents

As we learned earlier, most of Earth is covered by oceans. These waters are in constant motion, and the great currents that circulate through all of our seas have a dramatic effect on our climate.

There are two main types of currents. The first are surface currents, which are driven primarily by wind. They bring colder water from the poles to lower latitudes and warm water from tropical regions to higher latitudes. These currents can circulate over entire ocean basins in giant rotating systems called gyres. In the northern hemisphere, these gyres move clockwise because of the west to east rotation of Earth.

Labrador Current

Alaska Current

Gulf Stream

Current Events

Wind-driven ocean currents play a major role in the shaping of Canada's climate. The Alaska Current runs northward along the Pacific coast, bringing moist air that leads to British Columbia's mild temperatures and frequent precipitation. In the Atlantic, the Labrador Current brings cold, dry air south from the Arctic, inflicting harsh winters on the Maritimes. When the Labrador Current collides with the warm, moist Gulf Stream, the result is the region's famous fog.

In Canada, our coasts rub shoulders with two important currents. The first is the Alaska Current, which runs up our Pacific coast. Currents moving poleward are storehouses of warm, moist air, and when they occur on the eastern boundary of an ocean, prevailing winds carry the moisture over the adjacent land. This usually means warmer weather and more precipitation. That helps to explain why much of British Columbia has such a mild climate, and why the province has Canada's only rainforests.

On the other side of the country, the Labrador Current originates in the Arctic Ocean and heads down our east coast and around Newfoundland. This cold current is one of the reasons that Maritimers suffer much harsher winters than folks in British Columbia and helps to explain why our Atlantic coast has so many lighthouses. Off southeast Newfoundland, the Labrador Current meets the northbound Gulf Stream, which is really the primary weather engine in the Atlantic provinces. It brings warm water up from the Gulf of Mexico. The clashing of these currents creates the heavy fog for which our East Coast is famous.

There's a second type of current that occurs in the deep ocean. Cold water, like

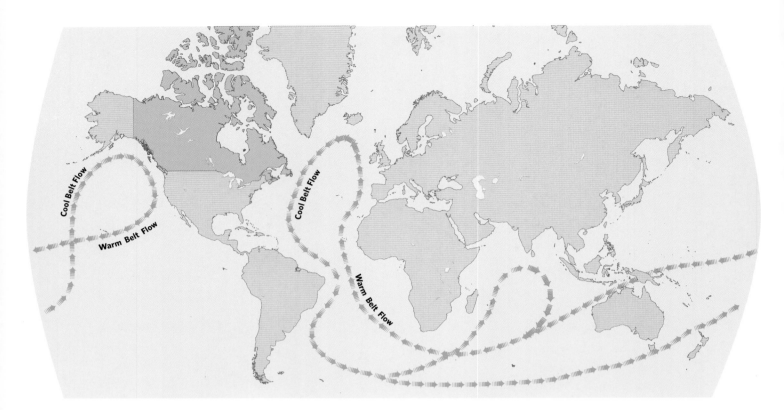

Great Ocean Conveyor Belt

The sun's energy is absorbed and redistributed around the planet by a complex system called the Great Ocean Conveyor Belt. The system consists of surface currents, driven largely by wind, and deep currents driven by differences in the density of seawater. If climate trends continue, global warming may actually lead to cooler temperatures in the North Atlantic, which in turn could cause winters in eastern Canada and western Europe to become colder by an average of 5 degrees.

The term nimbus refers to any cloud that brings rain.

cold air, is more dense, so it sinks. The density of seawater also increases with the amount of salt it contains. Just as differences in atmospheric pressure cause winds, masses of water with different densities generate currents in the ocean. These deep-water currents, sometimes called the Great Ocean Conveyor Belt, are extremely slow moving, but they transfer enormous amounts of heat around the globe.

Countries such as Germany and France are much warmer than Canada, even though they are at about the same latitude as we

are. The warm Gulf Stream is the main reason for this, and it gets a hand from the Great Ocean Conveyor Belt, which carries cold water southward from the North Atlantic. This cold water gets warmed up before returning north as a surface current along the coast of western Europe.

Oceanographers still don't understand all the effects deep-water currents have on climate, but they worry that global warming may slow them down. If that happens, both western Europe and eastern Canada will likely become much colder.

The Gulf Stream current off the coast of Newfoundland and Labrador is clearly visible in this photo taken from the space shuttle Endeavour.

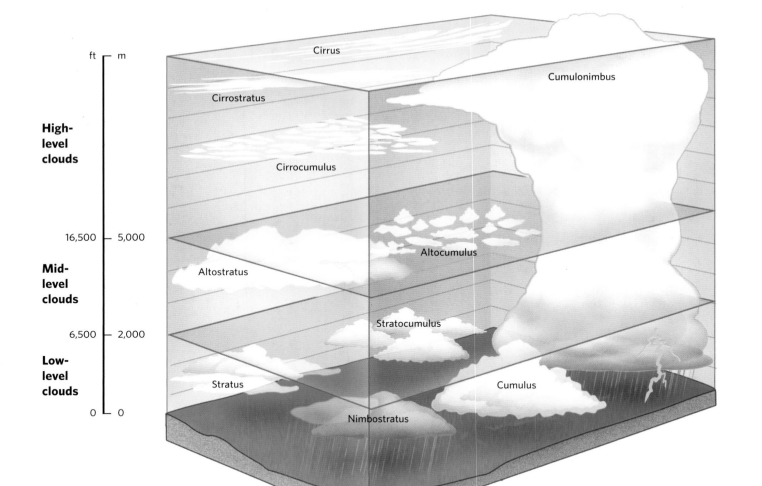

ft	m		
			Cirrus
			Cumulonimbus
High-level clouds		Cirrostratus	
		Cirrocumulus	
16,500	5,000		Altocumulus
Mid-level clouds		Altostratus	
6,500	2,000		Stratocumulus
Low-level clouds		Stratus	Cumulus
0	0		Nimbostratus

Cloud Types

Meteorologists classify clouds into several categories, according to characteristics such as altitude and shape. Cirrus clouds are the highest at 5,000+ metres, while stratus clouds are the lowest at below 2,000 metres. Those with the prefix alto- are mid-level clouds. A cumulus cloud is one with vertical development, while nimbus clouds carry potential for rain.

Q Which Canadian city is driest?

A On an average of 271 days each year, Medicine Hat, Alberta, experiences absolutely no rain, mist, fog, dew, frost or snow.

How Clouds Are Formed

Convection

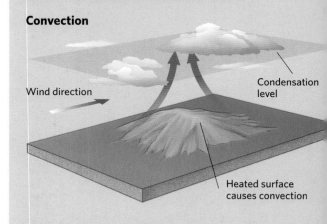

Wind direction

Condensation level

Heated surface causes convection

Convection

Convection refers to the movement of warm air from the surface to higher altitudes, where it condenses. Convection clouds can reach high levels of vertical development and may even extend the entire height of the troposphere.

Clouds

Do you remember when you were young and would stretch out on the grass, staring up at the clouds sailing by? There were all kinds of shapes and forms in the sky, and sometimes the clouds seemed almost close enough to touch.

What we didn't know then was that clouds are made up of little drops of water and ice—nothing more.

The first ingredient needed to form clouds is warm, rising air. This can be driven by convection, which occurs when air near Earth's surface is heated and becomes less dense than the air above it. It can also be caused by a process called orographic lift, which occurs when warm, moist air slowly drifts over a rising terrain and is forced into cooler air aloft. Warm air can also be pushed upward when it runs into a cold front.

As the warm air rises and encounters cooler temperatures, water vapour condenses to form droplets or (especially higher in the atmosphere) ice crystals. Minute particles of dust in the atmosphere attract these water molecules. As billions of these tiny droplets or crystals come together, they become the white or grey forms we call clouds.

Have you noticed that on a warm summer day, you can not only watch clouds form but also watch them fade away? When a cloud dissolves or vanishes, it's simply moving into an area of the atmosphere that is so dry it's evaporating away the droplets.

Clouds have been classified since the early 1800s. There are five basic types:

Roy Sullivan, an American park ranger, was struck by lightning seven times between 1942 and 1977. His hair was set on fire twice.

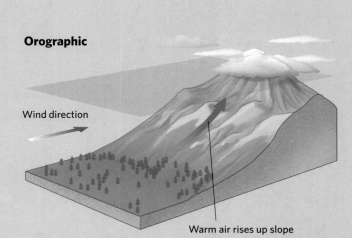

Orographic

Wind direction

Warm air rises up slope

Orographic Lift

Clouds can form when moist air is driven against a slope and forced upward, where it cools and condenses. The clouds that form during this process may be lenticular, or cap-shaped, and often remain stationary for many hours.

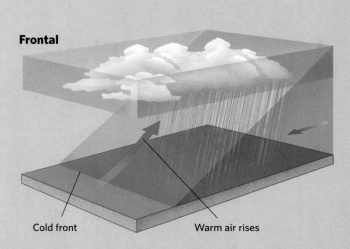

Frontal

Cold front

Warm air rises

Frontal Lifting

When moving warm air runs up against a cold front, the warmer, less dense air is forced aloft, where it condenses and leads to cloud formation.

Cirrus

Alto

Cumulus

Cirrus: Meaning "filament of hair," this term is associated with high-level clouds (above 5,000 metres).

Alto: Although the word is derived from the Latin word for "high," in meteorology these are mid-level clouds (between 2,000 and 5,000 metres above the surface)

Stratus: The term for low-level clouds (below 2,000 metres), it can also describe clouds that take on a layered appearance.

Cumulus: From a term meaning "piled or heaped," this label refers to clouds that take on a tall, vertical appearance.

Nimbus: Latin for rain, so in weather-speak, it's a cloud that carries rain.

These terms can be combined to describe more specific cloud types (see illustrations on previous page):

Cirrostratus: A layer of thin white cloud composed of ice crystals high in the atmosphere.

Cirrocumulus: High-level clouds

Stratus

Nimbus

composed of ice crystals that look like ripples or small puffs.

Altostratus: A layer of mid-level cloud composed primarily of water droplets organized in a uniform grey layer.

Altocumulus: Puffed grey and white clouds that have some vertical development and appear lumped together; they can also appear in waves at mid-levels of the atmosphere.

Cumulonimbus: Tall, rain-producing clouds associated with thunderstorms. They are puffy and rise high into the sky from a low, flat base.

Stratocumulus: Low-level, layered clouds with some vertical development that often take the appearance of large rolls or puffs which cover a large section of the sky.

Nimbostratus: Rain-bearing layered clouds that form in the lower and mid-levels of the atmosphere and cover the sky in a nearly uniform sheet of grey.

Precipitation

Precipitation is liquid water or ice that falls from the sky under the influence of gravity.

We noted earlier that clouds are composed of billions of water droplets or ice crystals. However, these particles are so small that the constant updrafts within the cloud structure keep them from falling. Some of the particles that do acquire enough mass to fall will end up in drier air surrounding the cloud and evaporate prior to reaching the surface. This failed precipitation is called **virga**.

In meteorology, we have classified what falls from clouds into three groups:

Liquid precipitation is rain and drizzle. You may not know that "drizzle" is a genuine meteorological term—it refers to water droplets, usually from stratus clouds, with a diameter between 0.25 and 0.5 millimetres. Rain, on the other hand, has drops that are larger than 0.5 millimetres, which fall primarily from cumulus or nimbostratus clouds. Sometimes fog is lumped in here, too—it consists of suspended droplets less of than 0.25 millimetres.

Solid precipitation is snow, ice pellets or hail. Snow is a series of ice crystals that are arranged in branches and hexagonal forms. No two snowflakes are alike, and the actual design of the flake depends on the moisture content and temperature. Meteorologists divide snowflakes into several types according to their appearance and the environment in which they are formed: needle flakes, column flakes, plate flakes, dendrites and star flakes.

Ice pellets are formed when a raindrop

The "smell" of impending rain comes from increased humidity, which intensifies odours.

Ominous thunderstorm clouds gather over downtown Toronto's harbour.

falls into sub-zero air and freezes to a solid mass. Hail begins as ice crystals at the top of towering thunderstorm clouds. The ice particles sink toward the earth, and as they do, they begin to melt or come into contact with liquid water. The strong updrafts often capture these particles and push them aloft again, where they refreeze.

This cycle continues, and the hailstone grows in layers. When the hailstones have grown too large to be supported by the updrafts, they plummet to the ground.

Both snow and ice pellets fall from cumulus and stratiform clouds, and Canadians obviously associate these forms of precipitation with winter. Hail, on the other hand, is most common in spring and summer, when thunderstorms are more likely to form.

Drop Zone

While illustrations usually depict raindrops with a tapered shape, they are actually spherical. Large raindrops typically flatten out due to air resistance, taking on an oblate shape. Raindrops less than half a millimetre in diameter, the size of the period at the end of this sentence, are classified as drizzle. Because of their smaller size, drizzle drops fall more slowly and are more susceptible to air currents, which can create mist conditions.

Raindrop

Drizzle drop

Spherical

Oblate

The frozen precipitation that fell during the 1998 central Canada ice storm caused massive damage to trees and electrical infrastructure.

1. Needle: –6°C to –10°C

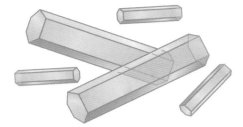

2. Column: –6°C to –10°C and below –22°C

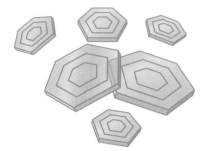

3. Plate: –10°C to –12°C and –16°C to –22°C

4. Dendrite: –12°C to –16°C

5. Star: –12°C to –16°C

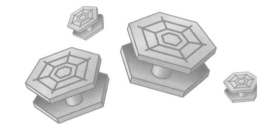

6. Column capped with plates: –16°C to –22°C

Frozen precipitation is the third variety, and it can be the most insidious, since it is responsible for creating very dangerous conditions. Freezing rain is a unique meteorological condition that occurs when a shallow layer of cold air (below 0°C) is in place at the surface, while a few hundred metres above it is much warmer—valleys are ideal for creating this situation. Rain falls as a liquid, passing through the thin, cold layer in just a few seconds, which isn't enough to cause the droplets to freeze. However, everything at the surface is below freezing, and as soon as the rain makes contact with the surface, it becomes solid ice.

Snowflakes

Every snowflake is unique, but snow crystals fall into general categories according to their six-sided shape. Each type forms under specific conditions and is influenced by cloud type, temperature and the density of water vapour.

Weather reporters in Canada almost never refer to "sleet." That's because this term is ambiguous—in Britain, sleet is a kind of wet snow, whereas in the United States, it means ice pellets. To avoid confusion, Environment Canada never uses the term, preferring "wet snow" or "ice pellets" instead.

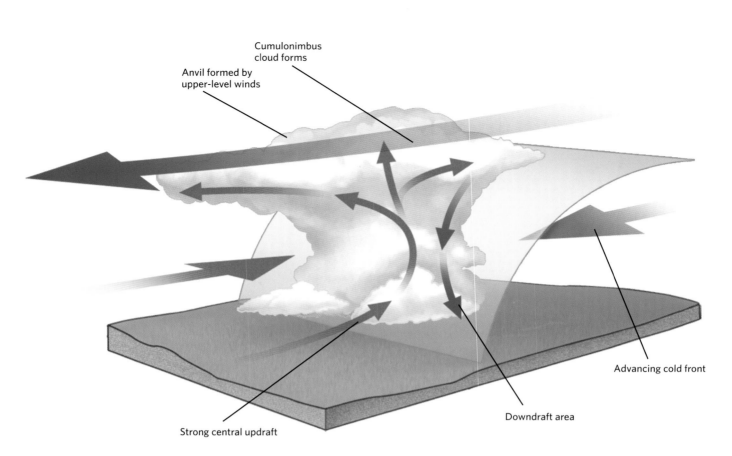

Cumulonimbus cloud forms

Anvil formed by upper-level winds

Advancing cold front

Strong central updraft

Downdraft area

How a Thunderstorm Forms

When warm, moist air is forced upward — in this case, by an advancing cold front — towering cumulus clouds begin to form, often taking on a distinctive anvil shape. Large amounts of moisture condense from the air; eventually so much ice and water form that the updrafts can no longer keep it aloft, and it falls as heavy rain. This increases the downdrafts of cold air, which ultimately cause the storm to dissipate.

Thunderstorms

Toronto's CN Tower is hit by lightning an average of 65 times a year.

Thunderstorms are spectacular and dangerous. Their beauty is in the towering clouds that soar up to the troposphere, sometimes exceeding 15 kilometres in altitude. The danger arises when these storms feature lightning, fierce winds, soaking rains, hail and even tornadoes.

For a thunderstorm to form, we need a moist air mass near the surface. A catalyst, such as convection or a cold front, is then needed to drive this moist air upward

in the atmosphere. If the temperature decreases rapidly with altitude, a thunderstorm may begin to incubate.

Thunderstorms have a three-stage life cycle. The first is called the cumulus phase: towering clouds grow and reach upward into colder air. During this first phase, the clouds fill with tonnes of water droplets, but updrafts prevent the water from falling.

The second phase is maturity. This is when ice particles develop in the upper reaches of the cloud. Eventually, enough ice will develop that the updrafts can no longer support their weight. At the same time, the upper air is becoming much colder and more turbulent. Water and ice begin to fall, increasing the flow of cold air toward the surface—this is the wind, rain and hail that are the thunderstorm's signature.

Cloud-to-air

Positive charge in upper cloud

Negative charge in air

Cloud-to-air lightning, usually the weakest form, occurs when a positively charged upper region of a cumulonimbus cloud interacts with a negatively charged area in the surrounding atmosphere.

Cloud-to-cloud

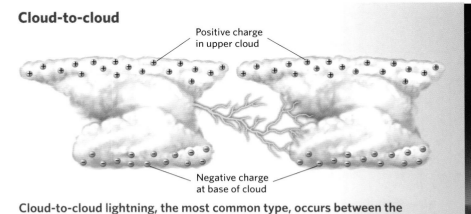

Positive charge in upper cloud

Negative charge at base of cloud

Cloud-to-cloud lightning, the most common type, occurs between the positively charged upper region of one cloud and the negatively charged lower region of an adjacent cloud.

Cloud-to-ground

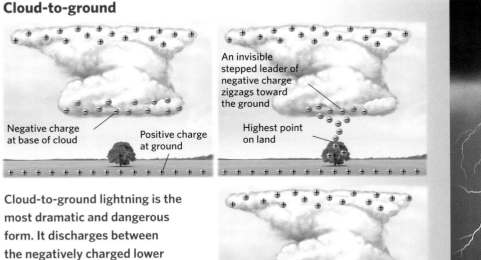

Negative charge at base of cloud

Positive charge at ground

An invisible stepped leader of negative charge zigzags toward the ground

Highest point on land

The circuit is completed in a lightning flash

Cloud-to-ground lightning is the most dramatic and dangerous form. It discharges between the negatively charged lower portion of a cloud and a positively charged area on the ground.

The final stage is the dissipation of the storm. The downdrafts of cold air and precipitation cool the air surrounding the storm, causing it to lose its fuel (moist, warm air). In time, the clouds evaporate in the cooler, drier air. However, there are times when a chain of storms will develop along a cold front. As the front moves into a large supply of moist warm air, it can give rise to a series of thunderstorms called a squall line.

Lightning is one of the hallmarks of a thunderstorm. It's caused when the upper part of a cloud takes on a positive electrical charge while the base becomes negative. (Scientists still aren't sure exactly why this happens.) Because air is a poor conductor of electricity, the charges will accumulate until the electrical difference is massive. Then the imbalance—remember, nature wants balance and harmony—is corrected by a huge discharge of electricity: a bolt of lightning. Thunder is the explosion of the superheated air surrounding the lightning, which can reach 30,000°C.

There are several types of lightning: cloud to air (that is, the lightning occurs within one cloud), cloud to cloud, and cloud to ground. In this last type, the strong negative charge at the base of a cloud responds to a strong positive charge on the Earth's surface. This why many people who've had a brush with lightning say their hair stood on end—that's the surface charge and cloud charge interacting.

Most tornadoes hit between 5 p.m. and 6 p.m., the fewest occur between 5 a.m. and 6 a.m.; in Canada, there are more tornadoes in June than in any other month.

Tornadoes

You may have seen movies about storm chasers—scientists who try to get as close as possible to tornadoes in order to study them. There are storm chasers all over North America who continue to investigate tornadoes, because we still don't know enough about how they form to accurately forecast where they will occur.

A tornado is a rapidly rotating column of air (a funnel cloud) that descends to the surface. Tornadoes are usually associated with massive thunderstorms called **supercells**. In these systems, the updraft that we see in all thunderstorms spirals to create what's known as a **mesocyclone**. When this vortex of air gets near the ground, condensed water and debris make it visible, often in the classic cone shape, but sometimes in a much broader wedge.

Tornadoes tend to develop from thunderstorms that form when dry and moist air are mixed rapidly, and Canada has two areas prone to this development. In Alberta, where there are no large bodies of water to add moisture, the humidity gets ferried over the Rockies from the Pacific. The province has lots of dry air that gets heated efficiently in summer, both by the sun and by the surface, which gets baked during the day. That hot, dry air rises rapidly and can meet with moist air flowing eastward in the upper atmosphere. This is an ideal environment for creating thunderstorms, and if the storm develops rapidly enough, those upper winds can move toward the surface and incubate tornadoes.

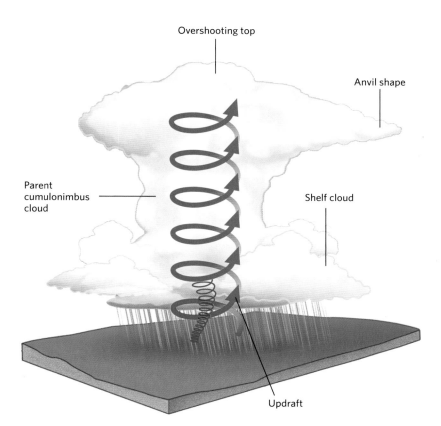

Overshooting top

Anvil shape

Parent
cumulonimbus
cloud

Shelf cloud

Updraft

Birth of a Tornado

Under the right conditions, a
supercell — a severe thunderstorm
with a strong, rotating updraft — can
give rise to a tornado. The spiralling
updraft, called a mesocyclone, can
be pushed toward the surface, where
its diameter becomes smaller and its
wind speed increases, forming the
tornado's characteristic funnel. One
indicator that a tornado is imminent
is the overshoot phenomenon,
whereby the flattened anvil shape of a
cumulonimbus cloud begins to bulge
at the top.

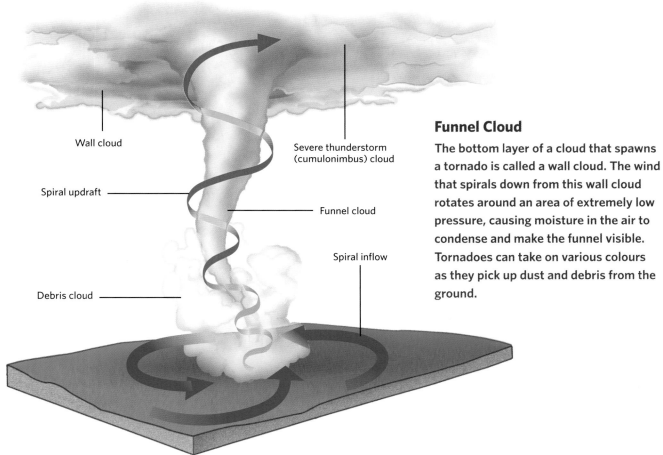

Wall cloud

Spiral updraft

Debris cloud

Severe thunderstorm
(cumulonimbus) cloud

Funnel cloud

Spiral inflow

Funnel Cloud

The bottom layer of a cloud that spawns
a tornado is called a wall cloud. The wind
that spirals down from this wall cloud
rotates around an area of extremely low
pressure, causing moisture in the air to
condense and make the funnel visible.
Tornadoes can take on various colours
as they pick up dust and debris from the
ground.

F5 tornado touches down in Elie, Manitoba.

Fujita Scale

No.	Speeds (km/h)	Damage Type
F0	64–117	Light (broken branches, damaged billboards)
F1	118–180	Moderate (moving cars pushed off course)
F2	181–251	Considerable (roofs torn off houses, large trees snapped)
F3	252–330	Severe (large trees uprooted, roofs and walls torn off)
F4	331–417	Devastating (houses levelled, cars thrown)
F5	> 417	Incredible (houses lifted and carried away, steel structures badly damaged)

The Great Lakes region is also suscept-ible to tornadoes, but for a different reason—the area is subject to what is called lake-breeze convergence. During the warmer months, the land is heated, the warm air rises, and then it's replaced by cooler air that blows in from the lakes. Look at a map of southwestern Ontario, and you'll see that much of it is nearly surrounded by lakes. The breezes that blow in from both Lake Huron and Lake Erie converge and—under the right conditions of high humidity and strong eastward upper winds—can create potent thunderstorms and possibly tornadoes.

The life span of a tornado is short when compared with other storms—weak ones can last seconds or minutes. Sometimes, however, they will form in chains or clusters, and these self-generating storms can go on for several hours before the atmosphere calms.

Tornadoes are measured on the Fujita scale according to their maximum wind speed and the damage they cause. An F0 tornado has winds of less than 115 km/h and can break tree branches, whereas the winds in an F5 exceed 417 km/h and can damage buildings made from reinforced concrete. The deadliest tornadoes in Canadian history—Regina's in 1912, and Edmonton's in 1987—reached their peaks at F4.

Tropic of Cancer

Hurricane Nursery

From June to November, the warm waters near the Tropic of Cancer (shown in orange) become a breeding ground for tropical cyclones. Rotating clockwise in the northern hemisphere, these begin as tropical depressions (areas of low pressure) and are upgraded to tropical storms as they grow. When winds reach about 120 km/h, the storms are categorized as hurricanes in the Atlantic. Similar storms in the northeastern Pacific — which are less likely to make landfall in North America — are called cyclones.

Hurricanes

It's rare for a hurricane to develop in Canadian waters, but it has happened. The only place in Atlantic Canada where the water ever gets warm enough for a hurricane to form is the Gulf Stream. Usually the hurricanes and tropical storms that we experience developed farther south and have drifted up to visit. It's also highly unusual for a hurricane to move more than a few hundred kilometres inland—although Toronto was hammered by the massive Hurricane Hazel in 1954.

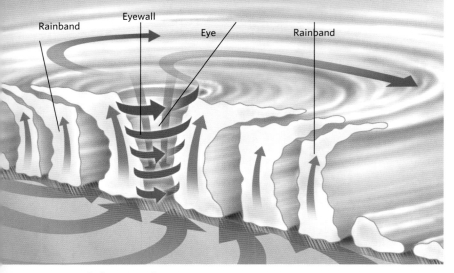

Rainband   Eyewall   Eye   Rainband

Eye of the Hurricane

A hurricane's eye is the area of very low pressure around which the storm rotates. The eye is an area of tranquillity when contrasted with the surrounding eyewall, the ring of cumulonimbus clouds surrounding the storm's centre that is the site of the storm's fiercest wind and rain. The line of thunderstorms that spiral out from the eye are called rainbands.

Tropical cyclones are named alphabetically as they occur, starting afresh at "A" with the beginning of each year. It is rare to reach "M" (13 cyclones) in a year.

A hurricane is a massive area of low pressure that forms in the tropics and usually covers an area at least 500 kilometres in diameter. These broad lows form all around the world at tropical latitudes and have different regional names. In the Atlantic and the eastern Pacific (adjacent to the Americas), we call them hurricanes. In the western Pacific, off Japan and China, they're called typhoons, while near India, they are cyclones.

The Atlantic Ocean becomes very active near the Tropic of Cancer in the summer, when sea surface temperatures are at their warmest—in some places the water is at 30°C—and the westerly airflow from Africa is strongest. A large area of high pressure becomes nearly dominant over the Atlantic Ocean between Bermuda and the Azores, and it sets up over the relatively cooler water north of the tropics. From here, the stage is set for hurricanes.

Hurricane season begins in June and runs to November. This is the time when all the necessary conditions are present. First, the water temperature must be at least 27°C to a depth of about 50 metres. The air above the warm water must also be reasonably humid. Finally, the winds aloft in the atmosphere must be steady in order to direct the storm and cause it to increase its energy. (If the upper wind speeds change or the air gets drier, the storm may not grow in strength.)

Low pressure areas begin as a tropical depressions; when the winds reach about 65 km/h and begin to swirl, the system is upgraded to a tropical storm. When the winds reach about 120 km/h, the storm becomes a full-blown hurricane. The life span of these storms—from their beginnings as a tropical depression until they dissipate—can be as long as 30 days. Every year about 80 of these cyclonic storms will form around the world, about 30 percent of which occur near the Americas, either in the Atlantic or the eastern Pacific.

You may have wondered who determines the names of hurricanes. Meteorologists at the National Hurricane Centre in Miami, Florida, have drawn up six rotating lists, usually using the names of friends or relatives. Storms are assigned names alphabetically, alternating between female and male, as soon as they are strong enough to be classified as tropical storms. In 2007, for example, the first tropical storm was Andrea, followed by Barry, Chantal, Dean, and so on.

When a storm is deemed to have caused significant hardship, its name is permanently retired. There will never be another Katrina, Mitch or Andrew.

The Saffir–Simpson Hurricane Scale

Category	Wind Speed (km/h)	Storm Surge (metres above normal)	Damage Type
1	119–153	1.2–1.6 m	Minimal
2	154–177	1.7–2.5 m	Moderate
3	178–209	2.6–3.7 m	Extensive
4	210–249	3.8–5.5 m	Extreme
5	>249	>5.5 m	Catastrophic

With gusts of wind exceeding 100 knots (185 km/h), Hurricane Humberto is shown here off the southeastern coast of Nova Scotia in 2001.

3

The Seasons

CANADIANS ARE THE beneficiaries of all four seasons. When you think about it from that perspective, there are many people in many places who aren't as fortunate as we are. There isn't a place in the country that doesn't see a snowfall each year. We all enjoy the warmth of the sun in the summer, and nearly every place in Canada has its share of autumn wind and spring rain.

The following pages will take us on a tour of Canada — every region, every time of year, a journey through vast and varied land coloured by the sunrises and sunsets of all the seasons. This is a trip that few of us ever really get to take. As we travel around the country and the calendar, we'll learn about the weather phenomena that are such an important part of the lives of Canadians.

The wonder of all this is that we see Canada in so many lights. Our landscape is shaped, redesigned and profoundly altered by each of the four seasons. We mark the passage of time as it is conducted through this annual cycle. The routine reminds us how powerful nature and weather are. It defines who we are in this land of varied seasons.

Top left: Butchart Gardens near Victoria, British Columbia, in spring. Top right: New Brunswick beach in summer. Bottom left: Lightning at Banff National Park, Alberta, in fall. Bottom right: Man shoveling snow, Cole Harbour, Nova Scotia, in winter.

Spring

S
PRING IS THE season that has the greatest impact on the optimism of Canadians. By late March, the sun has become stronger and the surface of the snow shines as the crystals melt and refreeze, forming a glaze of ice.

When the spring equinox arrives, the amount of daylight overtakes the hours of darkness and "solar loading" begins. This occurs when our atmosphere absorbs an increasing amount of energy as the northern hemisphere tilts more sharply toward the sun.

Alfred Russell Wallace (who with Charles Darwin co-founded the theory of evolution by natural selection) referred to our atmosphere as "the great aerial ocean." That's an appropriate description—after all, the atmosphere is vast and filled with currents and eddies, it undergoes great temperature fluctuations, and it never looks the same from one day to the next. Spring is when we see this aerial ocean come alive in its most dramatic fashion. Its actions begin slowly at first but steadily rise to offer the most fascinating and powerful storms known: tornadoes.

Our entire perspective changes when the spring thaw begins. Birds return, and animals become more visible as they search for mates. We, too, are gripped with spring fever—not an ailment but an instinct that tells us to go forward. There is a life force that comes from the strengthening glow of the sun.

Rainbow over cliffs, Magdalen Islands, Quebec.

Spring | The West Coast

Avalanches, such as this one at Mount Rundle, Alberta, can be a spectacular, and potentially deadly, event.

SPRING IN CANADA seems to arrive first on the West Coast—even by early March the temperature is rising. The snow that does fall is occurring high up in the mountains; the valleys are beginning to warm. Along the coast, both rain and snow are becoming less frequent, and the sun is shining a lot more.

What happens climatologically along the coast depends on how rapidly the large arctic high-pressure cell—which sits east of the Rockies—begins its retreat northward. As this cold air recedes, the jet stream also creeps farther north and a steadier flow of milder Pacific air begins to take hold.

Have a look at the average spring temperatures for these British Columbia communities:

Average Spring Temperatures

Victoria	10.7°C
Comox	10.3°C
Kitimat	8.3°C
Kelowna	10.1°C
Castlegar	10.6°C

Those temperatures look especially appealing when you compare them with provincial capitals farther east.

Average Spring Temperatures

Edmonton, AB	7.5°C
Regina, SK	6.9°C
Winnipeg, MB	6.7°C
Quebec City, QC	6.3°C
St. John's, NL	4.1°C

The increase in temperature afforded by more direct solar energy and a change in weather patterns can create problems as well. Avalanches are most frequent during early spring (although they occur with potentially devastating results in winter as well). An avalanche occurs when friction is unable to hold snow to the side of a slope. As snow accumulates on a mountainside, the terrain and vegetation help the snowpack adhere. As the amount of snow increases, its own weight compresses the pack. Meanwhile, the different types of snow—wet or dry, granular or flaked—determine how well the snowpack can maintain its integrity. Fluctuations in temperature, as well as the natural compression, ensure that layers of ice develop within the mass. The more layers that develop, the more susceptible the snow becomes to sliding from the mountain. All it takes is something to trigger it, and a heavy snowfall or rain is usually the culprit.

British Columbia has some lush forests, including the only rainforest in Canada. Along the north coast, exotic mushrooms and ferns grow under centuries-old fir trees whose diameters are measured in metres. All this because of the rain—there are places along the coast that see rain, or at least a misty fog, nearly every day of the year, not only in spring.

The interior valleys are drier. The southern Okanagan Valley is classified as

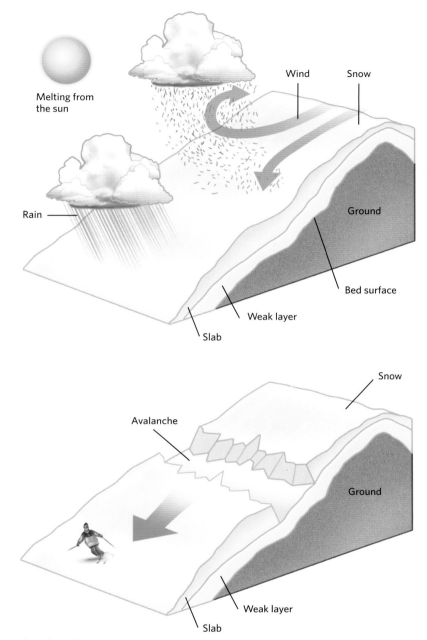

Avalanches

In the mountains of British Columbia and Alberta, skiers must be aware of the conditions that can lead to avalanches. When a heavy snowfall accumulates on top of an existing bed of well packed snow, the weak layer in between may be unstable. Rain on the lower slope may further reduce friction. Sometimes the accumulation of snow on the upper slope is so great that an avalanche may be triggered by a strong wind or even a loud noise.

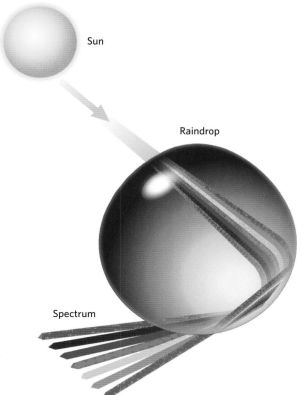

Sun

Raindrop

Spectrum

Nature's Prism

When white light from the sun passes through spherical raindrops, it is refracted (bent) and separated into its component colours: red, orange, yellow, green, blue, indigo and violet. We see this phenomenon as a rainbow, with the violet end of the spectrum on the inside of the arc. When a double rainbow is visible, the colours are reversed in the secondary bow.

One month's sunshine generates more energy than all the fossil fuels used by humans to date.

a desert—the cactus and sand variety. The rain and snowmelt that come in the spring are vital to sustaining the forests and farmland of these interior valleys.

Traditionally, April showers are said to bring May flowers. It's true in most places in Canada but not in British Columbia: by May, the flowers have been growing for a couple months, if not all year. During the last weeks of winter in Victoria, citizens count the flowers blooming in the capital and report the numbers—just to remind the rest of the country of what they're missing. In 2007, the tally was more than 3.3 billion flowers reaching toward the spring sun.

Rainbows, Halos and Sun Dogs

One of the most beautiful sights in nature is a rainbow framed by mountains. Rainbows are one of several displays that involve water and ice in our atmosphere.

Water droplets and ice crystals act like tiny prisms. When sunlight passes through the edge of a raindrop, it gets refracted (bent) and separates into its component colours: red, orange, yellow, green, blue, indigo and violet. These bands of coloured light are then reflected off the back of the drop, and because each colour emerges at a slightly different angle, we see them in distinctive bands. Red, which has the longest wavelength, is on the outside of the arc; violet, with the shortest wave-length, is on the inside. To see a rainbow, you must stand with the sun behind you.

The brightest rainbows appear when the sun is about 42° above the horizon, so you won't see one at sunrise or sunset.

Here's a neat fact: each raindrop presents only one colour. The angle at which you see the drop determines the colour you see. No two people can see the same exact same rainbow.

Haloes are caused by a similar principle. A halo is a rainbow-like circle or arc that appears around the sun when light is refracted by cirrus clouds at mid- to high altitudes. Cirrus clouds are wholly composed of tiny ice crystals, perfect little prisms for these displays. Sometimes you will see two particularly bright spots on opposite sides of the halo. These are called sun dogs or "mock suns."

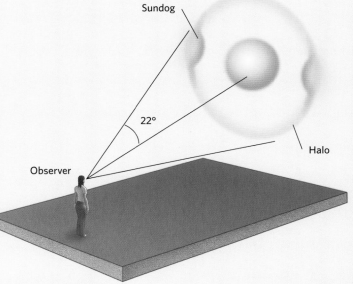

Halos and Sun Dogs

When the sun's light is refracted by tiny ice crystals in high-altitude cirrus clouds, the result can be a beautiful ring of coloured light called a halo. The type most often seen is the 22° halo, so named because this is the most common angle of refraction. Sometimes the halo includes particularly bright spots (sun dogs) on both sides of the sun.

Spring | The North

TO A NORTHERNER, spring comes at exactly the right time. To a transplant from the provinces, it seems to take longer, and the season is certainly not as prolonged as it is in the south. Here, north of 60°, you see spring before you feel it: you see it because the sun rises higher above the horizon and shadows grow shorter each day.

Spring is very short climatologically in Northern Canada and is marked by the return of migratory birds and the thawing of frozen land and water. Although new plant growth does not appear until nearly May, the increased strength of the sun is already apparent.

Here are the average temperatures for May in some Northern communities:

An avalanche near Chilkoot Pass, Yukon, buried 142 people and caused 43 deaths on April 3, 1898.

Average May Temperatures

Yellowknife, NT	5.6°C
Inuvik, NT	0.2°C
Hay River, NT	6.1°C
Baker Lake, NU	−5.8°C
Rankin Inlet, NU	−5.9°C
Iqaluit, NU	−4.4°C
Old Crow, YT	2.5°C
Dawson, YT	8.3°C
Whitehorse, YT	6.9°C

During spring weather, streams such as this one in the Northwest Territories carry off the melt water from the mountain snow.

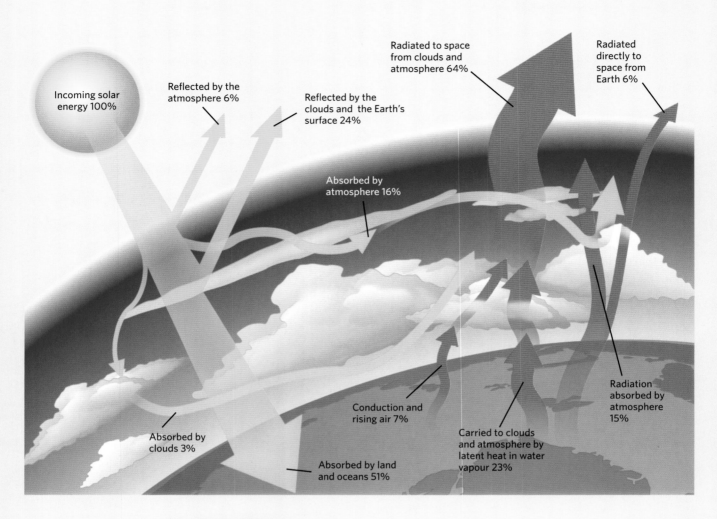

Incoming solar energy 100%

Reflected by the atmosphere 6%

Reflected by the clouds and the Earth's surface 24%

Radiated to space from clouds and atmosphere 64%

Radiated directly to space from Earth 6%

Absorbed by atmosphere 16%

Absorbed by clouds 3%

Conduction and rising air 7%

Carried to clouds and atmosphere by latent heat in water vapour 23%

Radiation absorbed by atmosphere 15%

Absorbed by land and oceans 51%

The Prime Mover

Solar radiation is our planet's ultimate energy source, and heat from the sun is the primary force behind our weather. The solar energy that reaches Earth (yellow arrows) is absorbed and reflected in complex ways by the planet's surface and the gases in the atmosphere. Radiation is also sent into space from Earth (red arrows). The atmosphere stabilizes the temperature on the surface by acting as a reservoir of solar energy.

Solar Loading

As the sun climbs high in the sky and offers the northern hemisphere more direct light and energy, an effect called solar loading begins to take place. The energy of the sun heats the atmosphere and the surface of the land, and this energy is slowly absorbed by our planet.

The transfer of energy from one molecule to another results in heat being created. This heat gradually radiates enough energy to melt the snow and ice of winter. You can begin to feel the effects of solar loading in early March: you'll notice it on a sunny but cool day, when your car's interior gets warm without your having to use the heater.

It is interesting to note that the communities with a May average that is still below zero are located in Nunavut. These places remain the coldest in the spring for two reasons. First, they are farthest from the milder airflow from the Pacific Ocean. Second, they are all closer to, if not abutting, the still frozen Arctic Ocean and Hudson Bay.

There are many ways to get around in northern Canada. Sled dogs and snowmobiles spring to mind right away, but air is the most common method. There are roads that traverse the Arctic as well, and northerners use open waterways during the short warm months. The roads tend to follow the waterways and ferries across rivers and lakes. These links are vital to the transportation of goods to and from the north. When the rivers and lakes are frozen, they become highways for transport trucks bringing all the items that people need: food, furniture, building supplies, fuel and everything else we take for granted south of 60°.

In spring, when the ice that covers rivers, lakes and tundra melts, there is a long interruption in the overland transportation system. The ice roads become unusable for as long as a few months during the spring thaw. Patience and planning are vital to northern life.

There are communities throughout the Arctic archipelago that rely on shipping lanes for supplies. It wasn't until 1969 that a large supertanker was able to transit the Northwest Passage through Canada's Arctic waters. The effort and energy required to move the Manhattan through this waterway was extraordinary: icebreakers kept working a path through sea ice and pack ice to ensure the ship's safe passage.

Today, freighters ply the northern ocean during the late spring, summer and early autumn months, providing essential service to northern towns. As our climate warms, the Arctic Ocean stays ice-free a little longer. This seems to help humans—at least in the short term—but it's having devastating effects on wildlife. Natural cycles are falling into disarray, and the habitat of polar bears, seals and many aquatic species is changing faster than the animals can adapt.

Spring arrives slowly in the north, but when it takes hold in mid-May, things happen rapidly. The snow and ice seem to vanish, and suddenly the landscape is painted in colour as lichens and small flowers blossom, taking advantage of the short window that nature has provided for them.

Snow accounts for 5 percent of precipitation worldwide; in Canada, that figure is 36 percent.

The Prairies

Farmers on the Prairies watch carefully for rain clouds like these.

AS THE WINTER snows melt away from the great rolling grasslands of our western provinces, what was once white becomes brown. Soon the strengthening sunlight and the right amount of rain will create a patchwork of green.

The spring rains are vital to those who make their livelihood from the grain belt. Science has allowed us to engineer species of wheat and other grains to better suit western growing conditions, but crops still depend on the rains of spring. When nature provides the required moisture, we receive a bounty of the world's best grains. If there is too much rain too early in the season, these crops can rot or develop fungus. If there is not enough, growth will be stunted and the harvest much diminished.

Prairie farmers constantly watch the spring skies. They watch for powerful storms—thunder, tornadoes, even the May snowstorms that visit like clockwork each year. The great display of all weather types occurs each spring because of the central position of the Prairies in relation to the rest of the continent.

As the sun warms the northern hemisphere, the entire weather pattern begins to change. The temperature rises at the surface and at mid-levels of the

Hail

Cedoux, Saskatchewan, has the dubious honour of receiving Canada's heaviest known hailstone, weighing 290 grams, which fell on August 27, 1973.

Thunderstorms feature powerful updrafts, or winds that circulate higher and higher into the atmosphere. For every action, there is an opposite reaction, so the storms also have corresponding downdrafts that circulate from high above us.

Hail forms when raindrops are caught in an updraft and carried to higher altitudes where they freeze. The static electricity created by these droplets and pellets bouncing around in the atmosphere draws them toward each other. The pellets grow until their own weight or a downdraft sends them toward the surface.

In very powerful storms, such as those in the Prairie spring, the growing pellets may be caught in another updraft and recirculated aloft, where they attract more droplets of water and further increase in size. Ultimately, the hailstones will grow so large that the updrafts can no longer support them, and gravity will have its way.

If you cut a hailstone in half, you will be able to see the onion-like layers, and by counting them, you'll know how many times the stone travelled through the storm. A hailstone the size of a golf ball would require 10 billion water droplets, and it would take the storm about 10 minutes to create. The largest hailstone in Canada fell near Cedoux, Saskatchewan in August 1973. The size of a grapefruit, it tipped the scale at 290 grams.

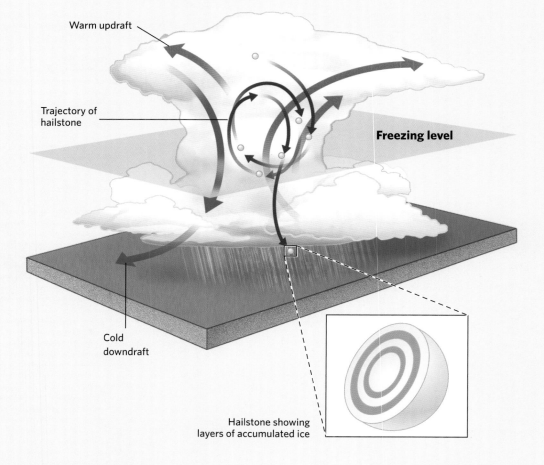

Warm updraft

Trajectory of hailstone

Freezing level

Cold downdraft

Hailstone showing layers of accumulated ice

atmosphere, and this allows two things to occur. First, the dominant high pressure that has been sitting over the west through much of the winter begins to move farther north. Remember, high pressure is simply cold and stable air settling toward the surface. With this retreat, the polar jet stream also recedes northward, causing a change in storm tracks and moisture sources.

As spring breaks, moisture is carried over the Rocky Mountains by a stronger flow from the Pacific Ocean. At the same time, with stronger heating from the sun, more powerful low pressure centres form just east of the mountains. These lows help draw Pacific moisture eastward, and they're able to access moisture from the Gulf of Mexico, too. It is this change that provides the needed rain; any slight deviations can result in sudden and dramaic changes to the weather.

In the spring, a different variety of storm begins to occur: thunderstorms. These can occur in every part of the country and in all four seasons, but in spring, the ferocity of thunderstorms increases as temperatures rise. All thunderstorms have the potential to bring heavy rain and strong wind; the most severe can also produce hail and tornadoes.

In Alberta, the area between Edmonton, Vegreville, Calgary and Drumheller is known as Hailstone Alley. More hail falls here than anywhere else in Canada. This area is one-half to one kilometre above sea level, so the surface is closer to the cold air at higher altitudes. A good supply of cold air from aloft is a critical component of mighty thunderstorms.

East of the mountains, on the flatlands,

This Calgary snowstorm during the early hours of April 10, 2008, made it a difficult trek to school for many children.

Upsloping Snow

I love Calgary, but its weather is possibly the most difficult to predict consistently. This has everything to do with where Calgary is located. It is close enough to the mountains to be in the rain shadow — the area just east of a mountain that receives less rain than the west-facing slope. Yet Calgary often gets freak snowstorms because of what is referred to as "upsloping."

In spring, Calgary can have the most amazing changes in weather, from 20°C to 20 centimetres of snow in less than a day. It happens when strong high pressure from the north centres over northern Alberta. South of here, a powerful low-pressure centre develops, generally in Montana and sometimes as far south as Colorado.

The high pressure circulates cold arctic air into southern Alberta, while the low pressure wraps moisture over the cold air. When both airflows arrive at the mountains, they are forced aloft into even colder air. The result is heavy snow on the east side of the mountain range.

This can occur anywhere along the Rocky Mountains, and even on a micro scale on high ground farther east. During the summer, the same pattern can emerge, but because it's warmer, the result can be heavy rains.

floods are the biggest worry of the spring. When you look at a topographic map of the three Prairie provinces, you see that the land slopes toward the east and north. The North and South Saskatchewan Rivers, the Peace River, the Athabasca, Assiniboine, Red, Nelson and Churchill Rivers and their feeders all flow north or northeast to empty into the Arctic Ocean.

This drainage pattern, coupled with the fact that ice and snow melt first in the south, creates potential for enormous problems. The meltwater in the south adds great volumes to these rivers, which then rushes northward, where it meets with ice dams. The result is that flooding occurs in many low-lying areas every spring. This has been going on since the end of the last ice age. As a matter of fact, that ice age left a depression in the surface that is now the prairies of Saskatchewan and Manitoba. The flooding that has gone on for thousands of years has left tens of thousands of lakes and bogs in the north of these provinces.

The Red River flows north through Winnipeg on its way to empty into Hudson Bay. When melting is too rapid during the spring thaw, the still frozen lakes and rivers north of Winnipeg cannot handle the added volume of water, and flooding is a regular and dangerous result. In 1997, the flooding of the Red River created a lake that covered 2,500 square kilometres. It was dubbed the "Red Sea."

Southern Albertans were granted an extra two weeks to file their tax returns after a series of storms dumped 1.75 m of snow in the area between April 17 and 29, 1967.

The annual spring cresting of the Red River, in Winnipeg, Manitoba.

Ontario and Quebec

THE ARRIVAL OF spring in Ontario and Quebec is a two-stage process. It doesn't arrive as noted on the calendar; rather, it evolves into place. The first stage is rather sloppy, as the dregs of a long winter are revealed by melting snow, and storms make one wonder when winter will release its grip. The second stage of spring is far more enjoyable: budding leaves and greening grass.

I've lived in Ontario since 1980 and have seen a lot of springs arrive. I really do believe that it is the season that takes longest to assert itself, perhaps because it is so eagerly awaited. The prolonged arrival is also because spring is one of two "shoulder seasons" (the other is autumn) that offer a transition between extremes.

Ontario and Quebec are in a confluence zone. That means the weather here is governed by several patterns: although the region is nearly landlocked, there is influence from the Atlantic Ocean, the Gulf of Mexico and Hudson Bay. The climate is part continental and part marine. In any given spring, the weather pattern can favour any one of these influences. A warm, wet spring indicates a stronger flow from the Gulf of Mexico. A cool, damp spring shows that low pressure in the Atlantic Ocean is prevailing. Dry and cool signifies the dominance of a polar

More than 168,000 cubic metres of water pour over Niagara Falls every minute.

Sugar Shacks

The maple leaf is widely recognized as the national symbol of Canada. There are 10 species of maple tree that are native to Canada, with at least one native to each of the 10 provinces. Not only is the maple tree valued for its beauty and use as a building material, but it has also become revered for a very special product derived from its sap: maple syrup. In North America, maple syrup is the preferred topping for pancakes, waffles and French toast, but it is also used as a sweetener for doughnuts, cereal, candied fruits and vegetables, breads and beverages. The production of this delicious sweetener has an important role in our early history, and Canada still produces nearly 80 percent of the world supply today.

The earliest producers usually built a small camp to protect themselves from the elements when boiling maple sap, and by the early 19th century, the sugar shack, with its vented roof to allow steam to escape, served not only as the hub of production but as a social gathering place. Beginning in the 1980s, communities in northern Quebec, and later Ontario, began opening sugar shacks, or Cabane à Sucre, to the public; these maple farms focus on

The maple tree flourishes in Canada and we have a long history of tapping its sap to produce syrup and other products, much of which is exported around the world.

educating and entertaining visitors with traditional production methods. Besides demonstrations, festivities and horse-drawn sleigh rides through the snow, many also have restaurants featuring traditional meals inspired by maple syrup and stores selling their products. One highlight of sugar shacks is fresh maple taffy. Maple syrup is boiled until it reduces and thickens, when it's poured over

fresh clean snow and rolled onto a popsicle stick.

In Quebec, particularly, this season has become an important celebration of culture, a way to pass down traditional skills and an important source of revenue. Many festivals exist across the country, including the one held in Saint-Georges, Quebec, which attracts nearly 50,000 people every year. In 2000, the festival in Elmira, Ontario, was listed in the Guinness Book of World Records as the largest one-day maple syrup festival, with 66,529 people enjoying the food, drink, music, dancing and much more.

After being shown simple methods to tap, collect and boil maple sap by Native Americans, early European settlers further developed the techniques, introducing wooden buckets and iron cauldrons in the 18th century. Over time, maple syrup became a staple part of the settlers' diet and was valued as an energy source, since white sugar was scarce and expensive.

By the 1960s, maple syrup production had turned into a major business and was no longer commonly undertaken by farming families. To reduce the amount of labour needed, tubing systems and vacuum pumps were introduced to bring sap directly from the trees to the evaporator house. Other improvements include reverse osmosis, which removes approximately 75 percent of the water content prior to boiling, thus reducing energy consumption.

The most commonly tapped maple tree is the sugar maple, but black, red and silver maple trees can also be used. Sugar maple trees are usually at least 30 years old and have a minimum diameter of 30.5 cm before they are tapped. Quebec produces over 75 percent of the world's supply of maple syrup, totalling 24,661,958 litres in 2005.

In the fall, maple trees produce a starch that acts as a kind of anti-freeze for their roots during the winter. In early spring, when daytime temperatures are above freezing (0°C) and nighttime temperatures below, this change turns the starch

into sugar, which mixes with water drawn from the roots. The warmer temperatures cause pressure to develop within the tree and this, combined with gravity, causes the "sugar water," or sap, to flow from holes in the tree's surface.

During this spring period, producers drill 1 cm holes approximately 5 cm into the trunk. Depending on their diameter and other conditions, as many as three taps can be installed on a tree. If managed correctly, this tapping is harmless to the tree, and some have been tapped continuously for more than 150 years; a new hole must be drilled every year, however, because trees will naturally heal these wounds. Once the holes have been made, spouts are inserted and covered buckets are hung, or plastic tubing is connected. The tapping period averages between four and six weeks and usually begins during the end of February or early March; the season ends and maple sap ceases to flow when nighttime temperatures rise above freezing.

When it flows from the tree, maple sap is thin, colourless and barely sweet; the distinctive taste is only derived through boiling. It takes approximately 40 litres of sap (the amount a mature tree produces in a year) to produce one litre of concentrated syrup after the evaporating process. A series of chemical reactions occur during the boiling process that produces the characteristically "maple" taste and colour we are used to and enjoy today.

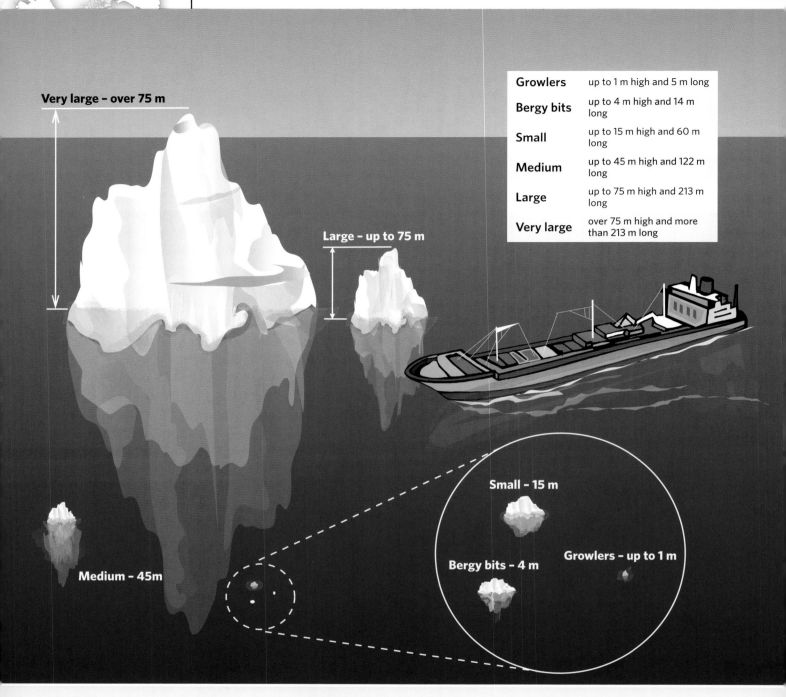

Very large – over 75 m

Large – up to 75 m

Growlers	up to 1 m high and 5 m long
Bergy bits	up to 4 m high and 14 m long
Small	up to 15 m high and 60 m long
Medium	up to 45 m high and 122 m long
Large	up to 75 m high and 213 m long
Very large	over 75 m high and more than 213 m long

Medium – 45m

Small – 15 m

Bergy bits – 4 m

Growlers – up to 1 m

Icebergs

In April 1912, the *Titanic* met with the North Atlantic's most famous iceberg about 600 kilometres southeast of St. John's, in the perilous Iceberg Alley. More than 1,500 people died, and about half of the 300 or so bodies that were recovered are buried in Halifax.

In Canada and elsewhere in the world, icebergs come from two sources. They are either great pans of sea ice that have broken away from an area of frozen ocean, or they are large sections of an ice sheet or glacier that have calved away from their parent formation.

These majestic islands of ice drift into the waters off Newfoundland and Labrador and Nova Scotia each spring, where they slowly melt. Some may get farther south, perhaps to the Gulf of Maine, but that's rare.

Because the density of pure ice is 920 kilograms

Tabular: Characterized by steep sides and a flat top.

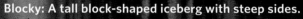

Blocky: A tall block-shaped iceberg with steep sides.

Drydock: An iceberg with a centre that has been eroded away to form a U shape.

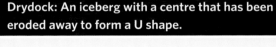

Pinnacle: An iceberg featuring one or more spires.

per cubic metre and the density of salt water is 1,025 kilograms per cubic metre, only about 10 percent of an iceberg is visible above the surface. It is what is beneath the water that poses a danger to marine navigation.

Icebergs are classified according to what is visible. (see illustration, facing page). They are also classified by shape: tabular bergs have a flat top with steep sides, like a plateau, and height-to-length ratio of at least one-to-five, while non-tabular icebergs can be domed or have a sloping surface. Some have spires or are eroded with channels.

An iceberg in the North Atlantic typically weighs between 100,000 and 200,000 metric tonnes. The largest ever recorded was more than 168 metres above the sea surface — that's as high as a 55-storey building.

One of the very few years in recorded history in which all five Great Lakes were completely frozen over was 1979.

Pond and Long Range Mountains, Codroy Valley, southwestern Newfoundland.

On April 13, 1984, an ice storm coated southeastern Newfoundland with 2.5 cm of ice.

are rooted in science: the old saying "red sky at night, sailor's delight, red sky in morning, sailor take warning" is correct about half the time. When the sun sets, the western horizon often glows red if there are clouds to refract the sun's rays. The idea is that the weather, which is moving from west to east, will pass by the time morning comes. A red sky in the morning indicates that the sun is rising through high-level clouds in the east; these clouds often signify that a low-pressure centre is approaching from the west and will bring foul weather that day.

The Labrador Current flows southward along the coast, carrying cold water from the Arctic Ocean to intermingle with the Atlantic Ocean near Newfoundland. This infusion of cold water helps create some interesting events in the east: thick fogs that can linger for days, strong storms and calm, clear days too.

The Labrador Current also offers a steady supply of icebergs every spring. These come from the ice that covers the Arctic Ocean, from the glaciers of Baffin, Devon and Ellesmere Islands or from Greenland. Many take years to inch from Baffin Bay into the Labrador Sea before finally reaching the waters of our Atlantic provinces. The Canadian Coast Guard does extensive monitoring of ice activity to ensure the safety of sea lanes. It may seem like a simple job to use ships, planes and satellites to follow the ice, but the process can get complicated when winds shift the direction of millions of tonnes of ice.

It always seems windy in Newfoundland. There is a stretch of road on the island that runs west and north from the ferry at Port aux Basques through the Codroy Valley, between Table Mountain and the Anguille Mountains. These peaks are a part of the Long Range chain, which runs along the west coast of the island. This area is home to the infamous Wreckhouse winds: the gusts here were said to be able to destroy homes, so no one would build there. In 1900, the wind was strong enough to blow over a locomotive on the Newfoundland Railway.

These winds are caused by the circulation pattern around a powerful area of low pressure and are enhanced by the terrain. If steep-sided valleys are aligned with the strongest prevailing winds, the gusts at the surface can be accelerated to more than 110 km/h. In the case of Wreckhouse winds, the valleys of southwestern Newfoundland are aligned perfectly with the prevailing storm winds from the southeast. The steep slopes funnel the wind and increase their speed. (In Nova Scotia—in the beautiful Margaree Valley on Cape Breton Island—similar southeast winds are called "les Suettes.")

In the early 1930s, the Newfoundland Railway heard of a local named Lockie MacDougall who had an innate ability to read the weather and always knew when the Wreckhouse winds would blow. The railway hired Lockie to warn of impending winds so the trains could be held. Legend has it that he was ignored by an engineer only once— and 22 cars on that train were blown from the rails. MacDougall performed this service for over 30 years and safely delayed the trains hundreds of times, surely saving countless lives. Today, signs warn transport trucks and minivans

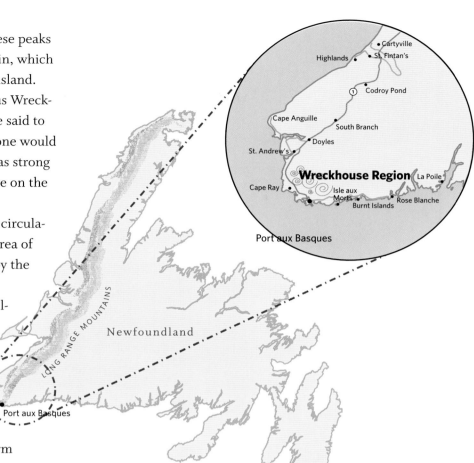

Wreckhouse Winds

The prevailing winds and topography of the Wreckhouse region of southwestern Newfoundland combine to create one of the gustiest places in Canada. When winds from the southeast run up against the Long Range Mountains, they accelerate as they are funnelled through a series of valleys. The resulting gusts are so powerful that they can blow vehicles off the road and strip the roofs from homes and barns. In 1900, the Wreckhouse winds faced off against a train and blew the entire train completely off the track.

about the perils of the Wreckhouse winds, but a plaque honouring Lockie still stands in Port aux Basques.

When the winds finally ease and the ice eventually disappears, a more pleasant side of spring shows itself in the east. In the Annapolis Valley of Nova Scotia, fragrant apple blossoms scent the air, welcoming the new season that's just around the corner.

Summer

N O MATTER HOW cold it may get in winter, the warmth and light of the sun offer all Canadians the bounty of life for at least a couple of months each year. The season we know as summer is filled with some of our favourite pastimes and memories. It is a time to enjoy the outdoors, to commune with nature on more pleasant terms than those offered by the harshness of winter or the unpredictability of spring and autumn.

Everywhere we look, we can see the evidence of nature's life cycles at their busiest stage. Green leaves are flush with the process of photosynthesis, which creates oxygen and nourishment from carbon dioxide, water and sunlight. Fruit is on the vine, ripening in the sun. Animals are rearing their young, not yet needing to think about the next winter. Humans are not much different. We plan our holiday time around this season and use it to rest, recharge and connect with family, friends and nature.

In meteorology, summer is when the sun presents its maximum energy to the northern hemisphere. The sun is at its highest in the sky in summer; the hours of daylight are at their maximum, the warmth is greatest, and we are also witness to some extreme weather.

The summer weather pattern in Canada sees the farthest retreat northward of the arctic air mass. This cooler pool of air also becomes slightly maritime in nature, in that it carries moisture from the Arctic Ocean. The dominant polar jet stream also shifts farther to the north as it circulates around a much stronger area of high pressure centred over the Great Plains. This allows significantly more tropical Atlantic air to flow over eastern North America and warmer, drier continental air to settle over the Prairies.

Summer officially begins with the solstice in June, usually on June 21. This is when the sun reaches its apex in the sky and offers the northern hemisphere its longest day of the year. From the first day of summer, then, the days begin to grow shorter. But the accumulated energy that the sun has been providing the land, water and atmosphere is just now reaching its peak. It's rather like bringing a pot of water to a boil and then turning the heat down to medium: the water will bubble for a considerable time as it holds on to the energy.

As Canadians, we feel we have the right to opine on the weather — after all, we deal with enormous extremes, and summer offers plenty of things for us to talk about with friends and family. The lack of rain—or the abundance of rain—is always excellent fodder for a chat, especially if you are in the agriculture business. We also get our share of heat waves, extreme bouts of humidity and days when the thermometer doesn't seem to be telling the truth. I think all of this makes summer our most enjoyable season.

The West Coast | Summer

THE CLIMATE IN British Columbia is uniquely and spectacularly diverse, and in the summer—when it is driest and warmest—it performs a delicate dance between ideal and disaster.

British Columbia is generally considered to have a maritime-alpine climate. However, it also displays several smaller-scale climates during the summer months: there is the temperate rainforest of the north coast and Queen Charlotte Islands, a true desert in the southern Okanagan Valley at Osoyoos, and a shortened northern summer in the Peace River region.

In British Columbia, the dominant weather pattern during the summer stems from a strong area of high pressure that brings seemingly endless sunny, warm days. This pattern is broken by surges of moisture from the Pacific Ocean. Most of the low-pressure centres are circulated

Rainforest, British Columbia.

northward, around the western flank of this massive high-pressure area, toward the Queen Charlottes. It is this pattern that generates the regional temperate rainforest environment along the north coast. Mild, rainy days are the norm in towns like Masset, Prince Rupert and Ocean Falls.

This same pattern of strong high pressure also keeps the interior valleys dry and warm each summer. As we've discussed, high pressure is a stable mass of air that is slowly settling to the Earth's surface, and as this air sinks, it compresses and heats up. This action, when combined with the strong energy from the sun, helps ensure that the interior valleys of British Columbia consistently generate temperatures that are much warmer than almost anywhere else in the country.

The west-facing slopes of the Coast Mountains are the beneficiaries of a near-constant flow of marine air. As this moist air meets the topography, it rises and condenses, bringing rain. This leaves little moisture for the inland valleys, and it is here we find the only true cactus-bearing deserts in the country.

Look at the difference in temperature and rainfall between a few interior valley communities in the south and those along the B.C. coast and to the north:

Temperature and Rainfall

Place	30°C Days	Rain Days	Summer Rain
Oliver	30	7	77 mm
Osoyoos	30	6	61 mm
Lytton	28	5	64 mm
Kelowna	21	7	102 mm
Kamloops	24	6	99 mm
Vancouver	1	9	140 mm
Victoria	1	9	85 mm
Prince George	1	14	165 mm
Prince Rupert	0	29	500 mm

It's clear that these are distinctive climate zones: the interior valleys in the south are warm and dry, while the coast and the northern interior are much milder and wetter. Again, it's all about that big high-pressure ridge.

In the rainier parts of British Columbia, lightning storms are a common occurrence in summer. Meteorologists still don't fully understand lightning. We know that positive and negative electrical charges develop inside storm clouds and that the ground also carries positive and negative charges. But how exactly these charges form a stroke or bolt of lightning is not clear: the debate surrounds whether a stroke is initiated from the cloud or

The mercury soared to 44.4°C in Lillooet, British Columbia, on July 16, 1941.

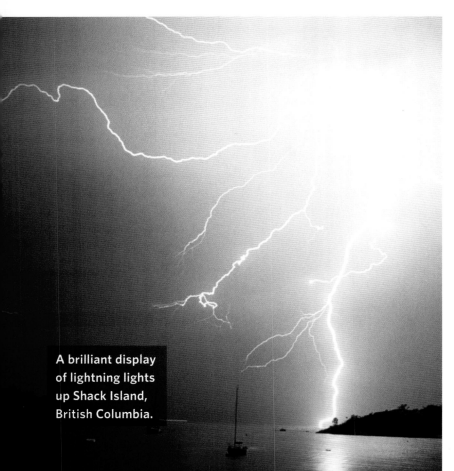

A brilliant display of lightning lights up Shack Island, British Columbia.

from the ground. Either way, we do know that the charge travels 96,000 kilometres per second (about one-third the speed of light) and that this connection can occur several times in one stroke, creating a flickering appearance. The temperature of an average lightning bolt often exceeds 22,000°C, and at this temperature, the air surrounding the charge is super-heated, causing it to rapidly expand and contract. This results in the creation of sound waves that we hear as thunder.

There is a weather rule of thumb that says by counting the seconds between when you see a flash of lightning and when you hear the thunder, you can estimate how distant the strike was. And it's true—it takes about three seconds for sound to travel one kilometre, so if you count the seconds from the flash until you hear thunder and divide by three, you learn how many kilometres away the lightning strike occurred. If the time between flash and thunderclap is getting shorter, the storm is moving toward you; if it is getting longer, the storm is headed away. As a general rule, you won't hear thunder occurring more than 30 kilometres away: the terrain and obstacles on the surface disrupt the sound waves enough to make them inaudible at that distance.

The warm climate in British Columbia's southern interior is nearly perfect for growing fruit and delicate vegetables. With the help of irrigation, farmers in this region have created one our nation's most prolific fruit baskets. The regular rains through the coastal mountain chain also afford us rich stands of forest—the Queen Charlotte Islands are rich with some of the Earth's most beautiful flora.

An incredible 25,000 hectares of forest were destroyed during the 2003 Okanagan Mountain Provincial Park fire.

Dry Thunderstorms

In hot, dry regions like the interior valleys of British Columbia, the rise of hot air and the circulation of warm air in the confines of a mountainous topography leads to strong electric charges in the atmosphere. All of that energy needs to be released, and thunderstorms often provide an outlet. These storms offer the usual mix of lightning, winds and thunder, but they often result in little or no rain because the air is so dry.

A simple lake breeze or slight vortex in the upper atmosphere is enough to trigger the development of one of these "dry thunderstorms." This kind of storm can be every forester's nightmare. The frequent strokes of lightning can start forest fires, and the lack of rain combined with the gusting winds creates ideal conditions for difficult to contain blazes.

That was the case in Okanagan Mountain Park and Kelowna during the summer of 2003. During one of the driest seasons on record, a lightning strike ignited a fire on the park's Rattlesnake Island in mid-August, and it swiftly spread. The fires burned for weeks and destroyed more than 25,000 hectares of forest, as well as 239 homes in southern Kelowna.

The North

T IS SIMPLY awesome to see how quickly summer explodes to life across our vast North. This nearly instant proliferation occurs because the sun rises higher in the sky and stays shining for extremely long periods of time each day. The indigenous life of the North uses all of this extra solar energy efficiently: plants grow rapidly, while animals and insects are on the move in great numbers, all relying on each other for the few short weeks of summer.

Imagine the sun slowly descending, but not actually setting below the horizon. During summer in the Far North, the days end with a prolonged dusk that gives way to dawn again without the darkness of night. Here's a comparison of the average daylight hours throughout the year in Inuvik and Toronto:

Average Daylight Hours

Month	Inuvik	Toronto
January	0:0	9:18
February	5:49	10:06
March	9:40	11:17
April	13:57	12:40
May	18:15	13:57
June	24:00	14:56
July	24:00	15:00
August	19:48	14:22
September	15:06	13:03
October	11:09	10:44
November	6:55	10:21
December	1:52	9:27

The tundra of the Ogilvie Mountains is beautiful at sunset, but it also plays an important role in controlling global warming.

The Tundra and Global Warming

The Arctic tundra is one of our planet's great carbon dioxide sinks. All plants remove carbon dioxide from the atmosphere as they photosynthesize during summer. When plants die and decay, they also emit CO_2, which contributes to global warming. But because of the permafrost just below the surface, and due to the rapid change from warm to cold seasons, arctic plants take a long time to decompose when they die. As a result, the tundra removes far more carbon dioxide from the atmosphere than it releases.

Now, as the planet warms, more and more tundra is thawing and staying thawed each decade. This means that the CO_2 that has been stored for thousands of years is now being released at a more rapid rate, accelerating the pace of global warming. Like the melting of ice — which leads to less reflected sunlight and even warmer temperatures — the thawing of the tundra creates a positive feedback loop: the increase in CO_2 emissions leads to warmer temperatures, which in turn leads to more thawing, and the cycle continues.

All of this additional sunlight plays a role in the stability of the weather across much of our North. With nearly even heating of the surface and atmosphere by the sun, there is far greater opportunity for stable high pressure to develop and remain intact. The farther away from the influence of the ocean, the more stable the atmosphere becomes. The strong high pressure delivers mainly sunny, dry days: the scattered northern weather stations report about one-third less rain than in southern Canada.

The treeline forms a natural boundary in the Arctic, separating the northern forest from the vast tundra. It extends from about Old Crow in the Yukon, eastward past the Mackenzie Delta to about the northern shore of Great Bear Lake. From there, it snakes southeast toward the Nunavut–Manitoba border on the shore of Hudson Bay. On the east side of Hudson Bay, the treeline bisects Quebec from La Grande River to Ungava Bay.

The hardy softwoods of the northern forest, mostly spruce and pine, decrease in size as you travel northward, until all that is left is scrub. Above the treeline, the climate and soil are able to sustain only smaller plant life. The region beyond is the famous arctic tundra, which encompasses most of the mainland north of the treeline. This massive peat bog is about a metre deep and is composed of slowly decaying plant life. Beneath the bog lies permanent ice—as the ice never melts, the swampy land cannot drain properly.

The Treeline

Because Arctic temperatures are too low to support large trees, the pines and spruces of the northern forest decrease in size as one travels north, eventually disappearing altogether. The boundary that separates the northern forest from the tundra is called the treeline, and it meanders around the globe, moving north and south according to local climate patterns. In Canada, it extends from near Old Crow, Yukon, to Ungava Bay in northern Quebec.

Treeline

The Dempster Highway, opened in 1979, is Canada's only all-weather road to cross the severe conditions of the Arctic Circle.

During the summer, the tundra is alive with plants and animals. As the surface thaws during the long summer days, it offers the moisture the plant life requires. It is an ideal breeding ground for mosquitoes and black flies too. During June and July, when it is warmest, there are literally clouds of insects searching for food. The tens of thousands of caribou that feed on tundra plants migrate through the Ivvavik and Vuntut National Parks to escape being eaten alive by the insect population. Farther north, on the massive and barren Arctic islands, the landscape of rock, sand and ice is amazing: black, brown and white as far as you can see.

In the Arctic Archipelago, the sea opens enough to allow navigation in the southern waters, but all of the Arctic Ocean, from the Queen Elizabeth Islands northward, remains a sheet of thick ice. These cold waters cannot generate significant low-pressure centres, so high pressure prevails in the northern summer.

I know several pilots who have flown in the North. They work seemingly endless days, flying into camps and centres of industry to deliver supplies for the winter. The roads are open too. The world's most northerly all-season road is the 671-kilometre Dempster Highway, which runs from Dawson City to Inuvik. It is named for William John Dempster, a corporal in the RCMP who regularly patrolled the area from Dawson City to Fort McPherson on the Mackenzie River. The only gas station lies near the midpoint, at Eagle Flats. The highway is dusty, barren and spectacular.

With high pressure offering generally clear skies and the sun staying up most all the time, the northern summer can be surprisingly hot. With few exceptions, the temperature does not usually fall more than 10 to 15 degrees from the warmest time of day to coolest, as you can see below:

Temperatures

Place	Avg. High	Avg. Low	Record High
Whitehorse, YT	19°C	6°C	34.4°C
Dawson, YT	21°C	6°C	34.7°C
Yellowknife, NT	19°C	10°C	32.5°C
Inuvik, NT	18°C	6°C	32.8°C
Rankin Inlet, NU	12°C	3°C	30.5°C
Resolute, NU	4°C	-1°C	18.3°C
Iqaluit, NU	10°C	2°C	25.8°C

The North is a barometer that tells us how our climate—and in a sense, our planet—is changing. Because of the dramatic differences between winter and summer, we can gain insight into the larger-scale changes that our planet is undergoing.

The Prairies

WHEN YOU THINK about the Prairie summer, what comes to mind? Visions of rolling wheat fields turning from green to gold. Farmers in their combines leaving a cloud of dust as they make their way through Canada's breadbasket. Summer unfolds from the Rocky Mountains and over the lakes of northern Saskatchewan to the forests of central Manitoba—it's the slow evolution of a season, interspersed with the dramatic offerings of nature: lightning, thunder, rains and tornadoes. For westerners, this is summer.

Summer on our Prairies can be punishingly hot. The highest temperature ever recorded in Canada occurred near the centre of the continent in Yellowgrass, Saskatchewan. In this quiet farming hamlet, about 20 kilometres up the road from Weyburn, the mercury topped out at 45°C on July 5, 1937. Estevan, less than

The rich harvest of the Prairies feeds people around the world.

a hundred kilometres to the southeast of Yellowgrass, is the sunniest place in the country, averaging about 3,000 hours of nearly clear skies a year—that's about seven hours out of every 24, all year long. This happens because southern Saskatchewan is influenced by high pressure systems not only during the summer but also in the colder seasons.

Hot weather is common farther west, too, where the average daily high in the summer hovers around the 24°C mark. There are more instances of temperatures reaching to the high 30s and low 40s here on the Prairies than anywhere else in Canada.

There is a perception that it is dry here on the prairie, and that is partly true.

Most communities from Calgary eastward to Saskatchewan receive about 25 percent less rain than communities in southern Ontario and southern Quebec:

Precipitation in Millimetres

City	June	July	Aug.
Calgary, AB	80	68	59
Medicine Hat, AB	63	41	33
Kindersley, SK	63	55	43
Regina, SK	75	64	43
Saskatoon, SK	61	61	39
Estevan, SK	76	65	49
London, ON	87	82	85
Toronto, ON	74	74	80
Ottawa, ON	85	90	87
Montreal, QC	83	91	93

The rain is infrequent because of the dominant high pressure, but thunderstorms can produce heavy accumulations over a short period. When the rain comes in this fashion, we lose much of

There is a perpetual threat of drought in the Prairies.

the water to runoff, because it does not have a chance to absorb into the soil. We have become better at conserving the rain we are offered, however: reservoirs and holding pools have been built to take advantage of what falls from the sky.

Despite our best efforts, though, drought on the Prairies can be devastating. The famous Dustbowl drought of the 1930s was a product of several years of stagnant weather: the global weather pattern had conspired to bring strong high pressure over central North America for several years with few breaks, and the above-average temperatures and below-average rain and snow baked the Prairies dry.

The drought earned the name "Dustbowl" because of windstorms that carried away so much of the dry western topsoil. Years of dry weather and poor farming techniques—which included removal of native grass—left the soil prone to erosion. There are true stories of soil falling like snow across the Great Lakes basin

Parts of Manitoba and Alberta were slammed with hailstones the size of oranges on July 24, 1996; damage in Winnipeg and Calgary was estimated at $300 million.

The Prairie climate in the summer months can generate tornadoes, like this F5 that touched down in Elie, Manitoba.

Prairie Twisters

Sixty percent of all reported tornadoes in Canada take place on the southern Prairies. Scientists still haven't determined exactly what causes a tornado to form, but we know all of the necessary ingredients that must be present for their development — and the Prairies have all of them.

Almost all tornadoes are born of severe thunderstorms, which offer the ideal conditions: rapidly rising warm air, rapidly descending cold air and strong converging winds aloft.

The Prairie climate in summer is ripe for the formation of potent thunderstorms: very dry and hot air is present most days at the surface. Prior to a large-scale change in the weather pattern, brought on by intense low pressure, we typically see three things.

The first is the introduction of a strong, low-level, warm, moist air flow from the south.

The second is a strong mid- and upper-level jet stream wind with cold, dry air from the northwest. The final element is a weakening of the high pressure: as its strength erodes, the warm, moist air is able to flow more strongly into the area, and it can rise farther into the atmosphere. The drier, colder air from the opposite direction helps to push the warm, moist air even higher and more rapidly. The result is powerful hail- and tornado-producing storms.

Cold front ▨

Cool, dry air at high altitude ━━━

Warm front ━━━

Warm, moist air near surface ━━━

Tornado Conditions

Most tornadoes in Canada occur on the southern Prairies, where atmospheric conditions are most favourable to the development of these deadly events. In summer, warm, moist air from the Gulf of Mexico moves into the Prairies, where it interacts with colder, drier air from the direction of the Rocky Mountains. This combination, together with a ready supply of warm, dry air near the surface, can give rise to massive thunderstorms, some of which spawn tornadoes.

and winter snow being red because it was mixed with so much topsoil. This pattern occurs on about a 25- to 30-year cycle, as we've witnessed severe droughts in the 1950s and 1980s as well. The results are devastating to the land and to the people who make their living from agriculture.

Of course, not all Prairie summer weather is dry: the region is also home to some of the country's most powerful thunderstorms, filled with torrential rain, wind, hail and occasionally tornadoes.

The origin of these thunderstorms is the great heating afforded by the dominant

continental high pressure. This stable air mass is an almost permanent fixture during the summer due to the upper-atmospheric pattern that establishes itself in the northern hemisphere. The energy of the sun feeds the northern oceans and landmasses, and that heat determines the overall weather pattern. Only the most significant disturbances will interrupt the summer weather pattern: low-pressure centres can develop just east of the Rocky Mountains and briefly alter the complexion of the atmosphere, and in doing so, they bring stormy weather.

There is another reason why storms grow so powerful on the Prairies: the absence of large bodies of water. There are plenty of small lakes and rivers, but no large ones in the south-central region outside of Manitoba—and the lakes in Manitoba are shallow and warm. The cooling offered by a large body of water would help to temper the relentless heating of the land and atmosphere.

Ontario and Quebec | Summer

"IT'S NOT THE heat, it's the humidity." You hear that a lot in central Canada during the summer. In this part of the country, the culmination of a prolonged heating of the atmosphere and surface frequently results in oppressive humidity. For those who like a near tropical climate, then southern Quebec and Ontario offer that. But for those who are not fond of hot, sticky days, summer may seem too extreme here.

The humidity is a result of the prevailing summer weather pattern in North America. The cool water of the central Atlantic Ocean allows for the formation of a massive area of stable high pressure called the Bermuda High, which circulates tropical air into eastern North America. The second ingredient is the heating of the central continent and the development of fluctuating thermal low-pressure centres that enhance the circulation of that tropical air.

Summer in Quebec City.

Humidex for Relative Humidity (RH)

RH (%)	100	95	90	85	80	75	70	65
T (C°) 21	29	29	28	27	27	26	26	24
22	31	29	29	28	28	27	26	26
23	33	32	32	31	30	29	28	27
24	35	34	33	33	32	31	30	29
25	37	36	35	34	33	33	32	31
26	39	38	37	36	35	34	33	32
27	41	40	39	38	37	36	35	34
28	43	42	41	41	39	38	37	36
29	46	45	44	43	42	41	39	38
30	48	47	46	44	43	42	41	40
31	50	49	48	46	45	44	43	41
32	52	51	50	49	47	46	45	43
33	55	54	52	51	50	48	47	46
34	58	57	55	53	52	51	49	48
35		58	57	56	54	52	51	49
36			58	57	56	54	53	51
37					58	57	55	53
38							57	56

This surge of warm, moist air from the Gulf of Mexico and the Atlantic is drawn along a path that follows the Appalachian Mountains. Along this route, the atmosphere is modified by several factors: the cool and dry air from the north, the heating of this air mass by the ever-warming surface and the increasing warmth of the Great Lakes. All of these elements work together to create the summer climate of this part of Canada.

Humid weather can be awfully uncomfortable. People stay cool by perspiring: when sweat evaporates from our skin, it removes a small amount of heat from the surface of our body. When repeated dozens of times each minute, this process helps make us feel cooler. The evaporation process requires that the air around us be able to accept the moisture that our body is exuding. When it is humid, the air is already filled to near capacity with water vapour, so evaporation is inhibited. We keep sweating, but the perspiration is now acting like an insulating agent. Instead of being cooled, we feel increasingly warm.

Meteorologists use an instrument called

Humidex Scale

Humidex	Degree of Comfort
20–29	No discomfort
30–39	Some discomfort
40–45	Great discomfort; avoid exertion
46 and over	Dangerous; probable heatstroke

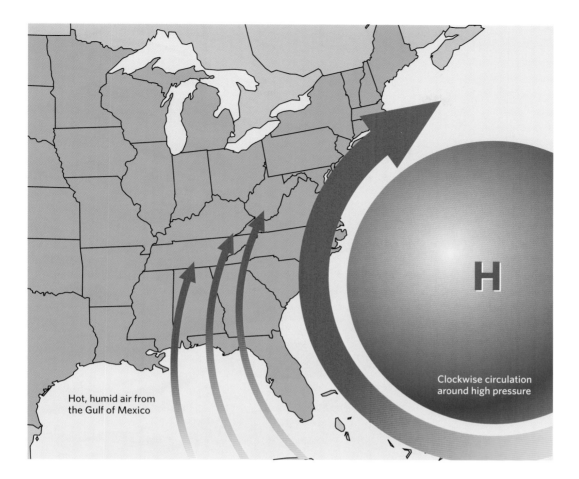

Hot, humid air from the Gulf of Mexico

H

Clockwise circulation around high pressure

Bermuda High

The oppressive summer humidity of Ontario and Quebec is due in large part to a prevailing system called the Bermuda High. This is an enormous region of high pressure over the Atlantic Ocean that is especially pronounced in summer. The system rotates clockwise, circulating warm, moist tropical air from the Gulf of Mexico into eastern North America, sending humidex readings soaring.

a hygrometer to measure the moisture content of the atmosphere. The key component of early hygrometers was human hair: the instrument measured the changes in the length of the hair caused by the moisture in the air. (Human hair actually becomes about 3 percent longer when it's humid.) You can use you own hair to get a reading on humidity: when it's muggy, curly hair gets curlier and straight hair goes limp.

In Canada, we refer to a "humidex," which reflects the added discomfort that humidity brings to hot summer days. A value between 30 and 39 makes most of us uncomfortable, while a reading between 40 and 45 comes with a warning to avoid exertion. Readings over 45 are dangerous. The highest humidex reading in Canada was recorded in Windsor, Ontario, in June 1953: it felt like 52°C.

Indeed, Windsor is the most humid city in Canada. It also gets the most thunderstorms, and this is no coincidence: there is a relationship between humidity, heat and storms. Warm, moist air rises, and

Canada's first recorded tornado hit the Niagara Peninsula on July 1, 1772.

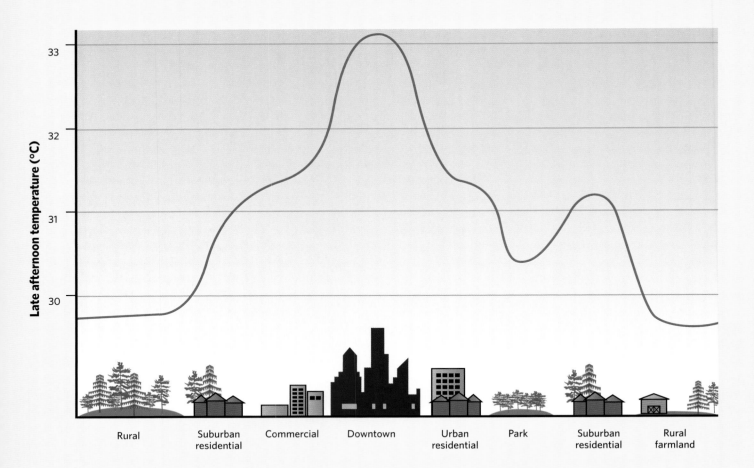

Heat Islands

As humans exert their influence on more and more of the planet, our affect on the climate continues to increase. One of the more interesting human additions to the weather patterns today is what is known as the urban heat island.

Large urban areas are often one to six degrees warmer at night than the surrounding countryside, and this is especially true during the summer. The replacement of vegetation with concrete, pavement and large buildings alters the way solar energy is absorbed during the day and released at night. These human-made elements have vastly different thermal properties than the natural landscape: they have a tendency to retain heat for much longer. During the day, the downtown core of many large cities will be a few degrees warmer than the suburban areas. Tall buildings reflect solar energy to the surface and alter natural wind patterns, curtailing the dispersal of accumulated heat.

At night, when the sky is clear, the day's heat is radiated back to the emptiness of space. But in an urban heat island, more of that heat lingers, and the artificial landscape stays warmer than nearby rural areas. This increase in heat creates changes in weather patterns: areas that are downwind of an urban heat island will experience higher amounts of precipitation due to the warming of the atmosphere over the city. This rising warm air condenses, and the rain falls just outside the city.

A hot summer day in Montreal, Quebec, where the effects of urban pollution are evident in the hazy sky.

when it is heated by the daytime sun, so-called cumulus congestus clouds may develop. These lead to the well-known "pop up" thunderstorms that regularly occur in summer.

Systematic frontal storms occur when dry cool air from the north arrives. This signals a large-scale change in the weather pattern that usually lasts for several days. The storms that occur along these large cold fronts will usually sweep through the entire region. The more dramatic the change in temperature and humidity, the more severe the storms will be.

All of this moisture in the atmosphere limits visibility, and summer forecasts often describe the weather as "hot, hazy and humid." The term "hazy" is a description of the sky: it usually refers to dust and mineral particles, such as salt, that are suspended in the atmosphere, giving it a yellow or brownish hue.

One beneficial aspect of haze is that it shields us from ultraviolet radiation. While UV rays are necessary for life, some wavelengths can be dangerous in high doses. Our natural defence against the rays is for our skin to release brown pigments—we see this release as a suntan. The amount of cloud in the sky offers some protection too, but in the summer, when we face most directly toward the sun, the intensity of these rays is highest. It is a misconception that UV rays are stronger on a hot day: their strength has nothing to do with the thermometer. The rays are strong on any clear day.

Q When and where was Canada's longest heat wave?

A Southern Manitoba and northwestern Ontario suffered the country's worst heat wave from July 5 to July 17, 1936. From Brandon to Dryden, daily temperatures approached 44°C. Approximately 1,180 people died in the heat wave, with many drowning while trying to stay cool in lakes.

The Atlantic Region

When I was growing up in the Maritimes, the summer always arrived as noted on the calendar, but it seemed to end during the last week of August. Now that I'm older and working with weather, that perception still appears close to how the season plays out in most parts of Atlantic Canada.

Summer in the east is spectacular. There can be very warm days, humidity and a few thunderstorms, but the dominant factor is a sea breeze that freshens and cools the air. The fact that nearly every place in Atlantic Canada is no more than 150 kilometres from the sea has a great effect: the ocean and its many arms govern the climate in this part of the country.

The summer weather pattern that tends to dominate eastern North America is strongly influenced by the Bermuda High that sets up in the central Atlantic Ocean. The western periphery of this area of stable atmosphere lies over the Atlantic provinces and affords light breezes from the southwest and brings generally clear skies. The humidity that plagues inland areas farther west is not such an issue here, because the cool ocean water precludes extended periods of very high temperatures. Look at these average summer temperatures for cities near to the coast and farther inland:

In Northwest River, Newfoundland and Labrador, thermometers reached 41.7°C on August 11, 1914.

Panmure Island,
Points East Scenic
Route, Prince
Edward Island.

Average Summer Temperatures (°C)

City	June	July	Aug.
Halifax, NS	19.4	22.9	23.0
Sydney, NS	18.9	23.0	22.7
Charlottetown, PEI	19.6	23.2	22.6
St. John's, NL	15.9	20.3	19.9
Moncton, NB	21.3	24.5	23.8
Fredericton, NS	22.8	25.6	24.7
Toronto, ON	23.7	26.8	25.6
Montreal, QC	23.6	26.2	24.8
Ottawa, ON	23.8	26.5	24.9

Argentia, Newfoundland and Labrador, averages 206 fog days per year.

The farther you are from the cooler waters of the ocean, the warmer it will be at these latitudes. This is important to the type of weather that occurs—and doesn't occur—in the Atlantic provinces.

In almost every part of the Atlantic region, the ocean is the great moderator of temperature. As the water continues to warm during the summer, it becomes an engine that powers the development of severe weather. The heating of the water alters the wind pattern over the course of

Peggys Cove, Nova Scotia.

the season, and all the available moisture makes for some very changeable weather.

Fog is my favourite type of weather. I grew up in the hills on the west side of Bedford Basin, in Rockingham, Nova Scotia. There were many summer days when a damp, cool fog would lay over the city all day. Other days would start out sunny and warm, and then around dinnertime, a great curtain of fog would drift in from the sea—it would literally roll up the street and envelop the neighbourhood in a matter of minutes. Visibility would drop to just 10 or 20 metres, and you could feel the temperature fall by five or six degrees. I also have vivid memories of driving to Wolfville from Halifax: the thickness of the fog would limit your speed and then, almost magically, you would break through to clear sky near Mount Uniacke.

Fog

Fog is simply clouds that form at or slightly above the surface, suspending millions of minute water droplets in the atmosphere. Fog occurs when the temperature changes cause water vapour in the air to condense. There are several varieties.

Radiation fog forms when the ground cools at night as the absorbed heat of the day radiates away. Clear nights are ideal, as there are no clouds to reflect heat back toward the surface. If only the air just above the surface is moist, the cooling will cause dew to form. If the layer of moist air at the surface is thicker — say, a hundred metres or so — fog forms. This type of fog is usually less than 300 or 400 metres from top to bottom and quickly dissipates when the sun rises.

Advection fog forms when warm, moist air drifts over cold water or cold land. Unlike radiation fog, advection fog can form during the daytime, and it actually moves and drifts over the landscape. It regularly occurs over large bodies of water — such as the Great Lakes — because water does not radiate heat in the same manner as land does. Advection fog is also shallow (about 300 metres) and can linger for an extended period when there is little atmospheric circulation. Often high pressure will hold this fog over an area for several days. Advection fog dissipates when the temperatures begin to harmonize or when a large-scale change in the weather pattern occurs, like the introduction of a cold front and dry air.

Valley fog is a type of advection fog that forms when air on the hillsides cools and then sinks into a valley in the evening, creating condensation. This type of fog can be slow to burn away in the morning.

Upslope fog occurs when moist air is forced up the side of a large hill or mountain. As the moist air rises, it meets cooler air and condenses, causing a fog that seems to hang on to the side of the terrain.

Autumn

OR A LONG time, I really didn't like autumn. I viewed it as the death of summer, with the first signals coming in late August or early September as the leaves began to change from green to yellow, red and brown. Now that I'm older, I see the true beauty of the season: cooler days after the heat of summer, longer nights to rest after the celebration of a season, an opportunity to reflect on what we've done and to imagine what we can accomplish next year. And all of this happens while nature entertains us with a final explosion of colour and a flurry of life before the onset of winter.

Depending on where you live in Canada, autumn can be a drawn-out and enjoyable transition period, or it can be a surprisingly short and dramatic prelude to a long winter's night. Many Canadians will tell you that the fall is absolutely the best time of year where they live, replete with harvest offerings and the signal of a new beginning. But there are just as many who see the season as an end to the vitality that the warmth of the sun nurtured over the summer months.

Fall begins with the autumnal equinox during the third week of September. After that celestial event, when the length of night catches up with and begins to surpass the day, the weather evolves rapidly. However, we have been on the road to fall since the very first day of summer. Following the summer solstice, the number of sunlit hours during each day begins to wane, and by the time autumn arrives, the amount of energy reaching the northern hemisphere is vastly diminished. Still, the lakes and oceans have built up a great reserve of solar heat, and the land also stores warmth from the spring and summer. In autumn, we find ourselves witnessing meteorological events that stem not only from the cooling of the atmosphere but also from the release of all that stored energy in the land and water.

Beginning in September, the oceans on our west and east coasts become the engines that drive the weather. By October, the water will generally be as warm or warmer than the land and the atmosphere. Currents generate storms and create waves of energy that will be transported by the atmosphere over the land. In the end, even the energy stored by the sea will dissipate, and by then, the season we call autumn will have ended.

We see the onset of fall in many colours across the country: on the West Coast, the arrival is viewed in white, as snow begins to cover more terrain on the mountains. The Prairies turn from gold to brown, and the great Canadian Shield is painted in orange, yellow and red as the leaves turn from their summer green. Even the colour of the sea changes a deeper blue, appearing almost black.

Across this country, the autumn offers a splash of brilliant colour before fading to the stark beauty of winter. It invites us to prepare for the long wait until the rays of the sun grow strong again.

The West Coast | Autumn

A LOT OF PEOPLE swear that fall is the most perfect time of year in British Columbia. It is true, in a sense: in the interior, the heat has eased, rain comes more regularly, and the season often provides day upon day of clear skies and pleasant temperatures. But we human beings have amazingly selective memories. The really nice weather is generally confined to late September and early October—after that, autumn becomes damp and chilly, no matter where you are in British Columbia.

As the northern hemisphere receives less direct sunlight, the atmosphere changes in complexion. The storm track shifts farther south as average temperatures over Canada begin declining. The change in both of these elements brings the cool autumn rain to the West Coast.

At this time of year, a semi-stationary low-pressure centre sets up in the Gulf

Waterfront houses at Deep Cove, North Vancouver.

Storm watching in Port Renfrew, British Columbia.

Average Temperatures and Precipitation in Coastal Versus Interior Locations

Place	Average	Sept.	Oct.	Nov.	Dec.
Bella Coola	Average temperature (°C)	15	8	3	0
	Average high/low (°C)	19/8	12/4	6/0	2/-2
	Precipitation (mm)	83	197	194	138
Sandspit	Average temperature (°C)	13	9	6	4
	Average high/low (°C)	16/10	12/6	8/3	6/1
	Precipitation (mm)	84	186	198	185
Campbell River	Average temperature (°C)	13	8	4	2
	Average high/low (°C)	19/7	13/3	8/1	5/-1
	Precipitation (mm)	59	153	231	215
Quesnel	Average temperature (°C)	12	5	-2	-7
	Average high/low (°C)	18/5	11/0	2/-5	-3/-11
	Precipitation (mm)	41	51	50	50
Castlegar	Average temperature (°C)	14	8	2	-2
	Average high/low (°C)	21/7	13/3	5/-1	0/-4
	Precipitation (mm)	42	50	93	92
Fort St. John	Average temperature (°C)	10	4	-7	-12
	Average high/low (°C)	15/5	8/0	-3/-10	-8/-16
	Precipitation (mm)	46	26	28	26

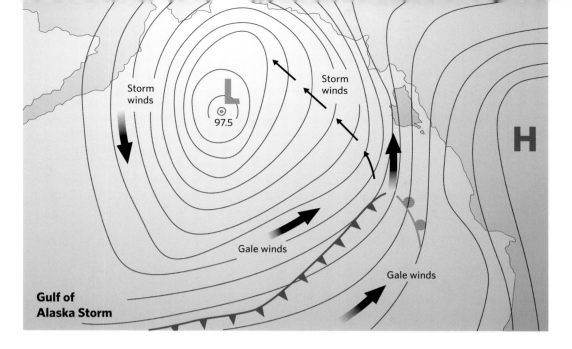

Storm winds

L
⊙
97.5

Storm winds

H

Gale winds

Gale winds

Gulf of Alaska Storm

of Alaska. Over the next several months, this feature tends to dictate the weather in British Columbia and the western Prairies. The so-called Aleutian Low oscillates in location and intensity during the fall and winter as it draws energy from the Alaska Current. This current circulates warm water from the North Pacific basin toward the coast of British Columbia and Alaska. All along the coast it becomes damp, but the warm water keeps the temperatures fairly stable, and it rarely approaches freezing.

The great surge of moisture and the rather consistent temperatures along the coast of British Columbia support the only rainforests north of the tropics. The pristine Gwaii Haanas National Park on the Queen Charlotte Islands is bathed in mist and rain year-round, but the heaviest rain arrives in the fall. This weather supports stands of trees that are centuries old and flora that makes it seem as if you have returned to the age of dinosaurs.

Just over the Coast Mountains, which form a natural barrier against the weather, the changes that autumn brings become

Aleutian Low

Early autumn weather on the West Coast can be pleasant, but by October, it can be damp and chilly throughout British Columbia. During the summer, the Pacific High, a stable system that brings fine weather, begins to weaken. It is replaced by the Aleutian Low, a prevailing low-pressure centre that forms over the Gulf of Alaska. By winter, this region can experience intense storms, which dump huge amounts of precipitation in coastal areas.

more apparent with each passing week. Shorter days mean less energy from the sun, and this equates with a cooler and more stable atmosphere.

Look at this comparison between communities that lie along or close to the coast and those that sit across the mountains in the interior (facing page). The effects of the warm ocean become clear.

The pattern is apparent: the Pacific Ocean provides a surge of moisture at the onset of the fall and at the same time moderates the temperature in coastal areas such as Bella Coola, Sandspit and Campbell River. In inland communities

The remnants of Typhoon Freda hit British Columbia's Lower Mainland on October 12, 1962, causing over $10 million in damage.

El Niño

A lot of mystery surrounds El Niño and La Niña. We know that these are large-scale climate events that occur in the tropical Pacific Ocean and that both involve changes in sea-surface and atmospheric temperatures. But the wide-ranging effect of these phenomena, which are felt around the world, are still not fully understood.

The name El Niño ("little boy") is a Spanish term for the Christ child; its name has been attached to the weather event because it is more noticeable near Christmas. The term was first documented in 1895, when a Peruvian navy captain named Camilo Carrilo described a pattern of warmer ocean temperatures off the west coast of South America. The warmer waters meant fewer fish and fewer seabirds, which the local people relied upon.

Sir Gilbert Thomas Walker, a noted British scientist, described the concept in 1923. Walker noticed that the warmer sea-surface temperatures associated with an El Niño event — which occur every two to seven years with varying strength — corresponded to a rise in air pressure over Indonesia, Australia and the southeastern Pacific, and a drop in pressure over the south-western Pacific. These periodic changes in the atmosphere are called the Southern Oscillation. El Niño episodes cause the trade winds to weaken, and the warm air rises along the west coast of South America, bringing plentiful rains. At the same time, warm water replaces the nutrient-rich cold water along the South American Pacific coast. The sea level in the eastern Pacific can rise as much as 60 centimetres over the course of this event.

El Niño alters global weather patterns enough that we in Canada generally experience winters that are milder by a few degrees, and on the West Coast, the spring and summer tend to be much drier.

La Niña ("little girl"), by contrast, refers to a cooling of sea-surface and atmospheric temperatures along the Pacific coast of equatorial South America, which leads to drier conditions. For Canadians, La Niña, offers the opposite of El Niño: cooler and wetter weather along the West Coast, and much drier and warmer conditions in central and eastern Canada.

such as Quesnel and Fort St. John, the temperature change becomes more dramatic and the moisture has less impact.

As we drift into the cooler months, the dew of summer becomes the first fall frost. This change in surface moisture signals that a change is underway. The valleys of British Columbia produce a variety of fruit and vegetables, but fortunately, early frost is rare: crop damage is more likely to come from hail or wind.

Once fall becomes established on our West Coast, we begin to experience some potent storms. These deep and powerful low-pressure centres are fuelled by the difference in temperature between the Pacific Ocean and the interior of the North American continent. As darkness descends over the Canadian north, the temperature of the continental air mass begins to drop rapidly. The colder it becomes inland, the stronger the West Coast storms may become. The Pacific lows deepen, and their winds increase; these November and December storms have wrought much damage through the years.

The coast of British Columbia lies exposed to the vast Pacific, and the airflow is not impeded as great pressure changes occur over vast stretches of the sea. The wind arrives on this coast—as it does on the Atlantic coast—with fantastic velocity. It is not uncommon to see storms with winds over 100 km/h arriving on the B.C. coast at least once a week from October through December. The orientation of the terrain further amplifies these hurricane-force gales, making our West Coast prone to some very dangerous weather during the fall.

The North | Autumn

FALL SEEMS TO arrive in Northern Canada even before the autumnal equinox. The daylight hours grow shorter much faster the farther north one travels. North of the Arctic Circle, the days of endless sunlight are gone by late August, and the nights become longer and longer, until 24-hour darkness arrives in November or December.

Even in Yellowknife, which never experiences 24-hour daylight, the nights grow longer at a much more rapid pace. Here is a comparison of daylight hours in Yellowknife (latitude 62.3°N) and Yarmouth, Nova Scotia (latitude 43.5°N):

Hours of Daylight

Date	Yellowknife	Yarmouth
August 15	16:05	14:00
August 30	14:35	13:19
September 15	13:00	12:32
September 30	11:30	11:47
October 15	10:00	11:03
October 30	8:30	10:20
November 15	7:00	9:40
November 30	5:45	9:11
December 15	5:00	8:56

Over each two-week span, the amount of daylight in Yellowknife decreases by about 90 minutes, while in Yarmouth the decline is only 40 minutes or so. The rate of the diminishing daylight has a dramatic bearing on the average temperature.

Autumn in the Richardson Mountains, Yukon.

You can see that the mercury in Yellow-knife falls about twice as fast as it does in Yarmouth:

Highs and Lows

Month	Yellowknife	Yarmouth
August high	18.2°C	21.0°C
August low	10.3°C	12.7°C
September high	10.3°C	17.8°C
September low	3.8°C	9.6°C
October high	1.0°C	13.1°C
October low	–4.0°C	5.1°C

The radical changes in daylight and temperature in the North create the potential for dangerous situations for anyone who has not arrived prepared. The poetry of Robert Service tells tales of the Klondike, and there are countless other stories of explorers and settlers whose fate changed as quickly as the season.

The animals must adapt quickly too. Those that migrate north to bear their young in the endless summer sun make a swift exit—by September, most migratory birds have begun their journey south, knowing that the increasing darkness means the climate will soon become inhospitable for them.

During the warm and light-filled summer, most Northern weather systems are formed near great bodies of water, and most are not overly powerful: for strong storms, we need greater differences in temperature and pressure. The Northern storms begin to grow more powerful as autumn arrives and temperatures begin to change in larger more rapid swings. It's all due to the diminishing level of energy from the sun.

The Earth's average surface temperature is 15°C. If Earth had no atmosphere, it would be –18°C.

The Franklin Expedition

One of the great mysteries of the North concerns the ill-fated expedition of Sir John Franklin. With two ships, HMS *Terror* and *Erebus*, and crews of about 130 men, Franklin set sail to navigate the Northwest Passage through Canadian Arctic waters in the spring of 1845. All of the crew eventually perished.

No one knows exactly what happened to Franklin's crew after they were last spotted off Greenland that July. Perhaps the arctic winter overtook them too quickly — the transition from summer to winter, without the usual autumn to which they were accustomed, may have proved too much for these explorers from temperate England.

Several search parties were commissioned to find the missing explorers, and some of these men also became victims of the North. Sir Edward Belcher sailed in 1852 to learn what had become of the expedition and the doomed search parties. On his multi-year voyage, he discovered much about where Franklin had travelled and found the remains of many of the lost.

Searchers discovered the locations where they set camp for their first winter. It appears that *Terror* and *Erebus* had become stranded in thick pack ice

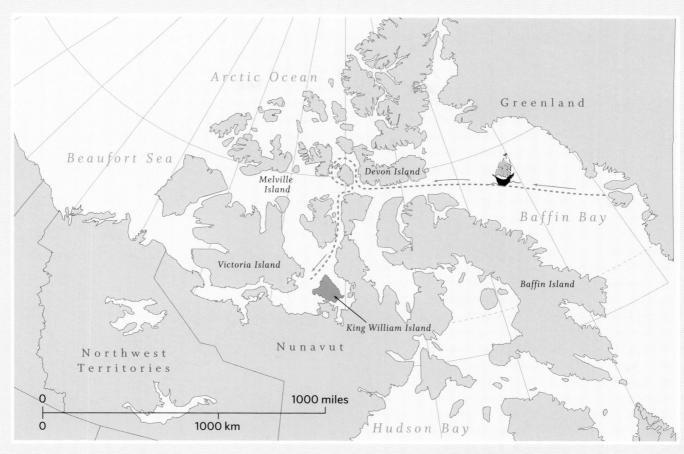

Franklin Expedition Route

and that the crew were forced to use the ships' timber for fuel. By 1847, Franklin and many of the expedition's members had died, and the survivors were forced to trek over land and ice, from King William Island to the mainland near the Boothia Peninsula.

Not one man survived the journey. They probably died from a combination of exposure, malnutrition, scurvy and perhaps even lead poisoning from poor-quality solder used in their canned food. There is evidence of cannibalism at some of the rocky gravesites that litter the route travelled by these aliens in a beautiful yet unforgiving land.

Grave markers from the Franklin Expedition.

An autumn storm
on the Thelon
River, Northwest
Territories.

Autumn | **The Prairies**

ACROSS THE VAST region we call the Prairies, fall arrives with a hue of gold as thousands of square kilometres of grain ripen. Clouds of dust rise in the wake of harvesters. The hardwood forests of Alberta, central Saskatchewan and Manitoba come alive with colour: the foliage takes on a new complexion as the sun's energy wanes and chlorophyll production slows.

In a good year, Prairie farmers will celebrate the harvest; in bad times, they'll make plans for surviving the coming winter and pray for a better crop next year. The autumn weather needs to co-operate so the offerings from the land can be taken.

In Alberta, Saskatchewan and Manitoba, the days remain long in the early fall, and farmers work until twilight to complete the harvest. Long ago, First Nations people knew the sequence of days as they read the land and sky; these wise people likewise harvested before the sun strayed too low on the horizon.

In the west, the weather pattern in the fall often provides fairly dry conditions — indeed, much of the rain that falls in summer comes from thunderstorms. As the atmosphere cools, those storms become less frequent, and rainfall diminishes sharply. The first frost usually arrives as the new season breaks, though most crops that grow here are not prone to its effects. As a matter of fact, some of the

The coldest Grey Cup football game ever, with a kickoff temperature of –17°C, was held on November 24, 1991.

Dry Island Buffalo
Jump National
Park, Alberta.

world's heartiest grain has been developed here: it is ideal for withstanding the first wave of cold.

The autumn does not offer much time to get ready for the harshness of the approaching winter. The early settlers learned in their first year here how quickly the seasons change. Those who came first and survived were lucky. This sample of monthly averages from southern and northern regions of our west shows how rapidly the cold moves in:

Monthly Average Temperature

City	Oct. High	Oct. Low	Nov. High	Nov. Low
Brandon, MB	10.8°C	-2.1°C	-1.1°C	-11.1°C
Thompson, MB	4.3°C	-4.3°C	-7.3°C	-16.6°C
Moose Jaw, SK	12.2°C	-0.6°C	1°C	-9.3°C
La Ronge, SK	7°C	-2°C	-4.4°C	-12.6°C
Medicine Hat, AB	14°C	-0.1°C	3.6°C	-8°C
Fort McMurray, AB	7.8°C	-2.2°C	-4.2°C	-12.8°C

By the end of October, the rivers and lakes are freezing over, especially in the north. Soon, the soil will begin to freeze as well. Above our heads in the atmosphere, the high pressure that brought warmth all summer has changed too. High pressure still tends to dominate, but the high that now lies overhead is arctic in nature rather than continental. In the fall, the polar air masses are sagging southward as the warm cell of continental air shrinks away from the mid-latitudes.

There will be brief periods when this encroaching pattern is interrupted, and these respites from the gathering cold are often followed by the first autumn snowfalls across the west.

Harvest Moon

The harvest moon occurs all across the country, but its romance — which features in so many Canadian songs and stories — is tied to the Prairies.

The harvest moon is simply the full moon that occurs closest to the autumnal equinox in the third week of September. Typically, the moon rises about 50 minutes later each night, but at northern latitudes in the fall, this period is only about 20 to 30 minutes. This means that when a full moon occurs near the equinox, there will be several days when the interval between sunset and moonrise is very short. Farmers in the past could take advantage of the extra light to harvest their crops.

It is also said that the harvest moon is bigger than other full moons, but this is an optical illusion. The size of the moon is always about the same, but when it is low on the horizon, we have terrestrial reference points that we can use to gauge its size. As it rises higher in the sky, our perspective changes, and the moon seems smaller. The full moon also takes on a reddish colour when it is low on the horizon. This is because we view the moon through more of our atmosphere, and most of the blue light is scattered, while the red wavelengths are visible. As the moon rises higher in the sky, we see it through less of the atmosphere, and the blue light is less scattered.

Ontario and Quebec | Autumn

ENTRAL CANADA IS witness to the arrival of fall in many forms. In northern Ontario and Quebec, it arrives quickly, as it does in the Arctic. Farther south, the onset of autumn is more subtle, marked at first by the change of colour in the hardwood stands, then by the final thunderstorms of the warm season, then frost that grows more widespread and harder each week, and finally, the snow that arrives and prepares the landscape for winter.

As the autumn settles in, a new climate pattern falls over Ontario and Quebec. The average temperature generally falls about eight degrees from month to month as the move toward winter continues unabated.

Laurentian Mountains, Quebec, in the fall.

Average Temperature

City	Oct. High	Oct. Low	Nov. High	Nov. Low
Windsor, ON	15.6°C	6.2°C	8.3°C	0.9°C
Toronto, ON	13.9°C	3.9°C	7°C	-0.7°C
Terrace Bay, ON	8.5°C	0.6°C	0.9°C	-6.5°C
Thunder Bay, ON	10.4°C	-0.5°C	1.7°C	-7.7°C
Quebec City, QC	10.7°C	1.7°C	2.9°C	-4.3°C
Val d'Or, QC	8.5°C	-0.5°C	0.1°C	-8.2°C

As in the Prairies and the interior valleys of British Columbia, agriculture is an important endeavour in the basin surrounding the Great Lakes and St. Lawrence River. A frost that comes early and unexpectedly can deliver a devastating blow to growers here.

The first signal that change is afoot is painted in the maple, birch, elm, oak and alders. As the last weeks of summer approach, the leaves become a cascade of orange, yellow and red, sweeping southward during the first weeks of fall until all the hardwoods have been denuded of their foliage. The landscape appears much less alive as the sun lies lower on the horizon, and the sky is grey

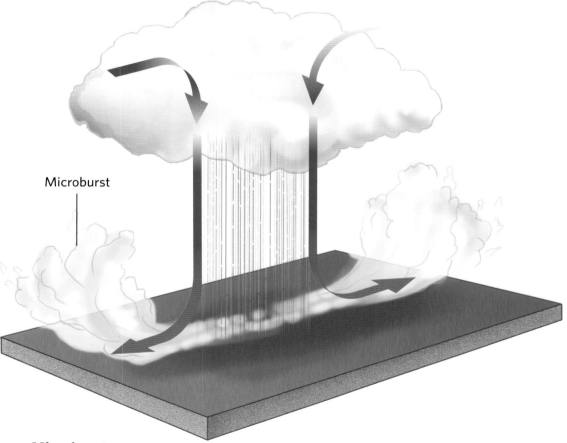

Microburst

Microbursts

While they last for only a few seconds, microbursts can generate wind speeds comparable to those in a hurricane. They occur when a thunderstorm generates an intense downdraft, which descends rapidly toward the surface. This massive column of fast-moving air is then deflected when it makes contact with the ground, generating gusts so powerful that they can pose a danger to aircraft.

more frequently, with low-slung clouds. The last migrations of birds sail overhead, moving south. The days are shorter, and the scent of frost hovers in the air with more regularity. The growing season comes to an end. All of this seems to happen quite suddenly, even though it has in reality been heading in this direction since the end of June.

During autumn nights, as the temperature declines, the air cools and condenses and the dew evolves into frost. The temperature at which the

Overlooking downtown Montreal from Mount Royal on an autumn afternoon.

condensation of moist air will force a droplet to occur is called the dew point. (This varies depending upon the moisture available, elevation, surface pressure and temperature.) When the dew point temperature is below zero, ice crystals rather than water drops form on these surfaces. These crystals of ice are frost. There are two types: hoarfrost is white and feathery, while rime is a much denser

Toronto was hit with its worst one-day snowfall on December 11, 1944, when 48 cm fell on the city.

Harvest Festivals

arvest festivals are celebrated around the world and have been for thousands of years; in North America, we've formalized this into the Thanksgiving holiday, a time set aside to give thanks for the crops collected during the autumn months. Few people today, however, can truly appreciate the enormous amount of effort it takes to grow and raise the food we find readily available at all times of the year in supermarkets. Canada is still one of the largest agricultural producers in the world, but due to improved technology and advances in science, the industry only employs

three percent of the population. Until fairly recently, however, most people relied on their farms for sustenance and were very much at the mercy of the elements and any natural disasters that might occur.

The lives of these Canadians centred on the changing of the seasons and how they affect the crops being grown. Harvest is the most critical period in the agriculture cycle as farmers must anticipate weather conditions such as rain, frost and unseasonably warm and cold periods; all of these determine crop yield and quality. As a result, farmers became the best predictors

of the weather, and publications such as *The Old Farmer's Almanac*, first published in 1792, continue that tradition today.

Harvesting has always been the most labour-intensive aspect of farming, so once that work was done, whole communities would gather together to celebrate and feast on the produce for which they had had to work and wait so long. In ancient times, various sacrifices were made to gods and goddesses, particularly those associated with fertility. These offerings, often in the form of burnt fruits, vegetables or cereals, were given in thanks and in hope that next

Pumpkins are a major part of harvest festivals.

year, they would be blessed with such bounty again. These early harvest rituals often included music, parades, sports and feasts.

Over the centuries, customs changed, but it was not until the mid-19th century that the hymns, special services and the practice of decorating churches with pumpkins, corn, wheat sheaves and other products of the harvest gained popularity. In Canada, the practice was brought over by European settlers and gradually evolved into the Thanksgiving holiday we now celebrate. The first Thanksgiving Day celebration after Confederation was "for the restoration of health to HRH the Prince of Wales" on Monday, April 15, 1872. It continued to be held on various days over the coming years and to be officially observed for different reasons. On January 31, 1957, it was finally decreed to be "for general thanksgiving to Almighty God for the blessings with which the people of Canada have been favoured," and a final proclamation fixed the date on the second Monday of October. Today, Thanksgiving is a statutory holiday in all jurisdictions except New Brunswick, Newfoundland and Labrador, and Prince Edward Island.

Over the years, the methods of celebrating and the focus of the occasion have continued to evolve. By the mid-20th century, Thanksgiving had grown primarily into an important social time when families got together to enjoy the fall weather and the autumn leaves, spend a final weekend at the cottage or play outdoor sports. Other celebrations, such as the Harvest Jazz and Blues Festival, which attracts 80,000 visitors to Fredericton, New Brunswick, focus on music, dancing and the arts, which have long been a part of the season. With fewer people closely connected to agriculture today, some churches and relief organizations are linking harvest time with awareness of those who continue to struggle to grow their food around the world.

Interestingly, it was not until the 1850s that roasted turkey became a staple part of harvest festivals. Most popular Thanksgiving dishes, such as sweet potatoes, corn on the cob, cranberry sauce and pumpkin pie, are made from foods that are native to North America. According to legend, these dishes were shared by the American Indians with the Pilgrims when they celebrated their first Thanksgiving at Plymouth in 1621.

The five largest sectors of agricultural production in Canada are grains, red meats, dairy, horticulture and poultry, which the country's temperate summers, ample grazing areas and fertile prairies make profitable. Farming methods continue to be improved, and weather conditions are predicted and anticipated with increasing accuracy. For these advances and our privileged position, Canadians should truly be thankful all year round, never forgetting the hardships early settlers went through to enjoy what we too often take for granted today.

Storm Surge

Storm surge, or storm tide, is a rise in sea-surface level that occurs during severe storms. It can cause massive erosion and flooding, making it one of the more devastating factors in hurricanes.

The increase of water level — which can be as much as several metres during powerful storms — is caused by several factors acting together. Wind is the catalyst: far away from land, the strong winds of a storm actually push or pile the water on top of itself, much like the way a

The rising waters of the Gulf of St. Lawrence lap at buildings on Tracadie Harbour Wharf, Prince Edward Island, during a 2001 storm.

wave rises when you push your hand through a pool of water. At the same time, the lower pressure exerted on the sea surface during a storm allows the water level to rise further. Finally, as the sea bed rises to meet the shore and water depth decreases, it creates a funnel effect that causes the bulge or wave to grow again.

idea of the magnitude of this current, it carries about 80 million cubic metres of water per second along the East Coast. By comparison, all rivers that empty into the Atlantic move at only 0.6 million cubic metres per second. The only current that is stronger circulates around Antarctica. The heat energy generated by the Gulf Stream is in the area of a hundred times the world's daily energy demand.

While we still don't understand exactly how cyclonic storms are formed, we are gleaning more information about the elements required for their development and growth. Warm water and steering winds are key ingredients. The sea-surface temperature must be at least 27°C to a depth of 50 metres: this amount of heat provides enough energy for convective activity to sustain itself over a prolonged period. The storms will only survive if they form at least 500 kilometres away from the equator. At this distance, the Coriolis effect can fortify the internal structure of the growing storm and allow it to become a weather system unto itself. There must also be pre-existing areas of disturbed weather with which the storm can interface. For example, strong high pressure over Bermuda helps circulate a storm toward the east coast of North America.

A tropical storm is upgraded to a hurricane when the winds reach 120 km/h. By this point, the centre will be a well-defined "eye," where the air pressure is significantly lower than the surrounding area. Severe hurricanes are categorized according to their wind speed and the height of their surge. As the colder northern air becomes more dominant over Atlantic Canada, the sea surface cools and is no longer able to support tropical storms and hurricanes. The late autumn storms that develop over the Maritimes and Newfoundland and Labrador, however, are as powerful as any hurricane, even if they don't carry that name. The interaction of cold air over the still warm water here allows for the incredibly rapid genesis of low pressure.

Indian Summer

Well into autumn, Atlantic Canada will usually have one more flirtation with summer. The period of sunny and warm weather that follows the first frost is often referred to as "Indian summer."

In Europe, a period of sunny and warm weather in autumn was often named for saints whose feast days occurred around the same time: "St. Luke's summer" was a warm spell around October 18, while "St. Martin's summer" occurred about November 11. By the 18th century, the European settlers who arrived on this continent began to call these warm autumn periods "Indian summer," though the origin of the term is shrouded in controversy.

Some people believe that settlers coined the term when they observed the aboriginal peoples using these periods to take in their harvest or to hunt game. Others suggest it arose because the warm weather was more pronounced in the territories formerly inhabited by aboriginal peoples than it was in the eastern colonies.

Gulf Stream

A young citizen of Prospect Village, Nova Scotia, explores the wreckage of the local wharves after Hurricane Juan's passage.

Winter

WHEN WE THINK of Canadian weather, it's usually winter that comes to mind.

Our winters can include darkness and mind-numbing cold for weeks on end, or they can bring nearly endless rain and temperatures that cause one to check the calendar to verify that it really is winter. It's the coldest and bleakest time of year in our country, and it illustrates the temerity of people who chose to live with and tolerate this harsh weather. But to many, it's also the most beautiful time of year, evoking images of a blanket of freshly fallen snow under a clear blue sky. Some Canadians happily insist that it's their favourite time of the year; others smile and accept the season, knowing that spring eventually follows.

I grew up in Halifax, one of the oldest cities in our country. I remember learning in school about the early settlers in this land, and I was always fascinated by the stories of their first winters here. Those who were fortunate enough to be close to the aboriginal people usually survived their first winter with fewer deaths. That's how harsh the winter is in Canada: it will kill you if you aren't prepared.

A few hundred years ago, for many, it was a triumph to simply make it through the season. But we soon became a part of it. Because we are a nation of strong, creative people, we embraced winter with open hearts. We invented hockey. We became great skiers, skaters and curlers. We built snowmobiles and snowblowers. We have adapted and flourished in the harshest of climates.

Back in 1947, Snag, a nearly inaccessible little town in the Yukon, recorded the coldest temperature ever in Canada: −63°C. That's still a ways off the coldest temperature ever recorded on the planet, which is −89°C at the Vostok science station in Antarctica, but once you're past 60 below you probably wouldn't notice the difference. When it's that cold you can actually hear a squeaking sound as your breath condenses as soon as you exhale.

As for snow, we get our share. Surprisingly, though, the high Arctic receives some of the least snow in all of Canada. This is because during the winter months, the Arctic Ocean is frozen solid, and there's no supply of moisture to generate snow. The Canadian Arctic is actually among the most arid places on Earth.

The most snow in an average winter falls in Gander, in Newfoundland and Labrador. This town on the shores of the Atlantic Ocean is also home to some of the country's strongest onshore winds and fiercest winter storms. Gander will often see more than four metres of snow fall in a single winter.

Cold and snow instantly define winter in Canada, but there is a lot more to our winter than those two faces. Follow me across Canada, and learn about the hidden surprises our winter offers.

WINTER IN BRITISH Columbia is defined for me by the fact that it arrives on time, in December, and leaves when it should, in March. It runs right on schedule. Very British.

The British Columbia winter is a season of contrasts. It may be wonderfully snowy in Whistler, but you'll need an umbrella in Vancouver, just a two-hour drive away. You are also very likely to be surprised by warm days in the mountains and, at least once a year, by a huge snowfall in Victoria. It's the only province where you can golf and ski on the same weekend in January.

Indeed, there are several distinct climatic zones in this vast and varied province. The coast remains temperate all year—the winter in Vancouver is similar to the winter in Prince Rupert, nearly 800 kilometres to the north. The reason is the warm Alaska current that flows from the southwest,

This Vancouver pedestrian shelters himself from February rains.

The Alaska Current

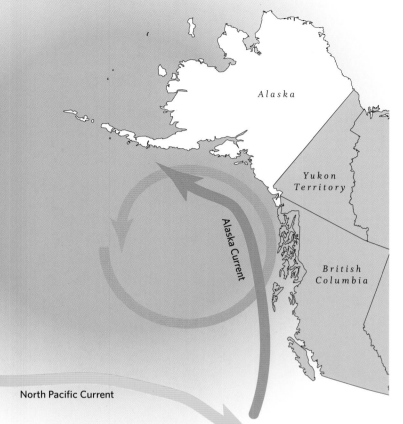

The Alaska current flows north.

North Pacific Current

The Pineapple Express

Some of British Columbia's heaviest winter rains and snowfalls are the result of warm, moist air carried all the way from Hawaii on the Pineapple Express.

During the winter, the subtropical jet stream — a high-altitude air current — steers warm, humid air from the tropics up to the west coast of North America. This air isn't just moist, it's saturated, which means it contains as much liquid as it's possible to carry. Saturated air quickly turns to precipitation when it encounters a small temperature fluctuation, and that's just what happens when it arrives on our coast. The air mass meets the mountains, which cause it to cool, and it dumps its load of rain or snow.

keeping the air above the water mild. A plentiful supply of moisture, together with a prevailing onshore wind and the mountainous topography result in weather conditions seen nowhere else in Canada.

Prince Rupert and other parts of the central coast region are temperate rain-forests, thick with vegetation that is fed by steady rain and moderate temperatures. This city receives precipitation nearly 20 days a month on average, with an annual total of over 2,500 millimetres: that's 100 inches of rain.

Of course, not every region of British Columbia is rainy in winter. The Coast Mountains serve as a barrier, keeping mild Pacific air from entering the interior valleys. Generally, winter temperatures are about 5 to 10 degrees colder in the interior than on the coast. The interior also gets nearly 25 percent less precipitation, and the bulk of that is snow.

Temperature almost always decreases with elevation—the rate is about 6°C for every 1,000 metres—so it's no wonder that the province's many mountains are beautifully snow-shrouded. The average winter temperature in Vancouver and Comox hovers around 4.5°C, but you only have to climb 750 metres to find temperatures below zero—cold enough for snow to fall and stay. Revelstoke, in the southeast of the province, had the greatest seasonal snowfall in Canada during the winter of 1971–72: more than 24 metres.

In the northern coastal valleys, winter is usually very temperate. Towns like Terrace and Stewart on the north coast are at the heads of deep fjords, where the scenery is stunning. Here the average winter temperature is easy to take—just a few degrees below zero—and there are more than 300 centimetres of snow in an average winter, or more than a metre a month. Canada's greatest one-day snowfall occurred on February 11, 1999, at Tahtsa Lake, just southeast of Terrace: 145 centimetres.

British Columbia's normally mild winters are occasionally interrupted by a phenomenon called arctic outflow. Several times each winter, a large high-pressure centre moves south from the Arctic, where it gets sandwiched between the Pacific coast and the Rocky Mountains. This brings vast quantities of dry and very cold air (as low as –20°C) into regions where the air is usually moist and warm.

British Columbia, of course, is a land of mountains and valleys, and during an arctic outflow, the cold air settles in the low-lying areas. As it circulates from the east, it is accelerated westward as it gets funnelled through the valleys—especially those that are oriented east-west—just as a river's current increases in narrower places. As wind speeds pick up, the temperature drops, and wind chills become fierce. The strongest winds are those that sweep through the fjords, where they're sometimes called squamishes.

What's worse, once the cold air settles into the valleys, it tends to stay a while. Remember, cold air is denser and harder to move, especially from a valley. It will eventually move when the sun warms the air enough, but in the winter, with less direct sunlight and shorter days, that takes longer to happen.

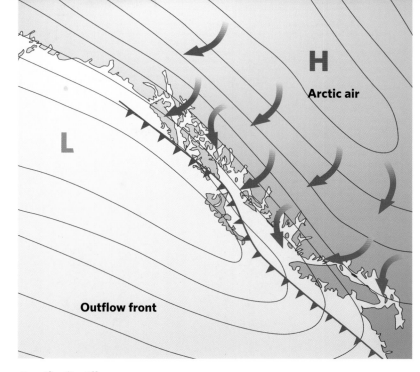

Arctic Outflow

During the winter, a mass of cold and dry air from the Arctic can descend on British Columbia, where it becomes wedged between the Rockies and the Pacific coast. This high-pressure centre can bring a spell of frigid weather to a region that usually enjoys mild winter temperatures.

The sight of Salmon Glacier, Stewart, British Columbia, and other glaciers like it, is increasingly drawing tourists from around the world.

Ice Roads

Canadians have many unique ways of turning extreme weather conditions to their advantage, and a prime example of this is the creation and use of ice roads. Tools, food and supplies were originally moved into remote Northern areas by dogsled, but by the 1950s, ice trails were being used to support motorized transport. Today, ice roads are the most practical means of access to remote areas in the Yukon, Northwest Territories and Nunavut.

Ice roads are surfaces of ice on large bodies of water that have been prepared to bear the weight of large tractor-trailers and other vehicles. Such roads are often composed of several such frozen waterways linked by overland portages. During some of the winter season, they allow for the relatively cheap transportation of goods to and from isolated regions that would normally only be accessible by plane or boat. Ice roads often provide the only practical means of moving building materials and heavy equipment for the many mining camps in the North.

Many Northern communities are unable to maintain long stretches of

A fuel transport truck on its way to the diamond mines of the Northwest Territories.

permanent road because of the high cost and the boggy condition of soil, known as muskeg, during the warmer months. For areas with no permanent road access, an ice road is a crucial transportation link; the Northwest Territories, for example, has 1,400 kilometres of ice roads and crossings, more than the length of all its paved roads combined. Accordingly, ice roads are heavily used during the season in which they are navigable, which, depending on the region and conditions, can be anywhere from a few weeks to a few months. Some mining camps bring up nearly a year's supply of materials during this time, running trucks along the ice roads 24 hours a day.

The ability of ice to support a load depends on a number of different factors, including its thickness, the pressure of water below it, the way the ice formed and anomalies in it, whether the body of water is salt water or fresh, and the speed of and distance between vehicles travelling on it. To prepare an ice road, a path is plowed across a frozen body of water; removing the insulating layer of snow exposes the ice to subfreezing temperatures as low as –51 °C, causing it to thicken. A system of auger holes is sometimes used to flood and further thicken the ice. Although ice 70 centimeters thick will usually support a light vehicle, a depth of at least 100 centimeters is needed for heavy transport trucks. Regular testing done with manual equipment and Subsurface Interface Radar determines the depth of the ice and whether conditions will support the weight of vehicles. During the spring thaw, an ice road is last to melt; even in the summertime, bare sections on the bottoms of lakes and rivers can often still be seen from bush planes, revealing where the thick ice prevented sunlight from stimulating plant life and algae.

Although ice roads allow goods to be moved economically, there remain many hazards to truckers hauling cargo. Subfreezing temperatures can easily cause hypothermia or frostbite, and winter storms can prevent aid from arriving. Other dangers include breaks in the ice, called pressure ridges, created when heat causes the ice to expand and contract. When a truck travels on an ice road, the pressure creates waves below the surface; if the pressure is too great, the ice can be blown out ahead of a vehicle, particularly just before land portages when the water gets shallower. To prevent such disasters, speeds on ice roads are generally limited to 24 km/h, and stopping is prohibited. It is also very important that ice roads are regularly plowed to prevent snow from insulating and warming the ice; if warming occurs, cracks and ruts may appear, possibly resulting in a vehicle falling through the ice. To allow for better chances of escape, most drivers do not wear seat belts.

Every ice road is unique, both from other ice roads and from season to season. One of the most famous, and the longest in the world, is the Tibbitt to Contwoyto Winter Road, which extends 568 kilometres, beginning 60 kilometres east of Yellowknife and ending at the Jericho Diamond Mine, on Contwoyto Lake, Nunavut. Like many ice roads, it is private and operated by a group of mining corporations to bring in their heavy equipment and supplies; diesel fuel is one of the most common loads. The road was first constructed in 1982 and has been rebuilt every year since; it is 50 metres wide with 87 percent of its length on frozen lakes and the rest over 64 land portages. It costs approximately $10 million to operate this road for the average of 67 days that it is in operation. In 2007, the ice road was open from January 27 to April 9, and a total of 10,922 truckloads were hauled across its length, resulting in only nine accidents and one minor injury.

In the summer months, it is hard to imagine that an ice road bears thousands of trucks across this stretch of water near Yellowknife.

Winter | The North

ON THE CALENDAR, winter begins in December. In the North, however, climatological winter has been under way for nearly two months. In Yellowknife, the average low is already below freezing in October.

Some people joke that there are only two seasons north of the 60th parallel: winter and blackflies. Another adage says there are two seasons: the one when you can get around and the one when you can't. Indeed, when you look at a map of our North and see how far-flung the communities are, you begin to realize the importance of transportation links. Most travel is accomplished by air or on the few roads that we've carved into the wilderness. If you drive in the North, you'll gain a true appreciation for our inventiveness. During the warm months, ferries cross the rivers. In the winter, ice roads follow the rivers. It can actually be easier and cheaper to move around the Arctic during the winter, though it's also far more dangerous.

Winter conditions vary throughout the Arctic because of its sheer size. What we refer to as the North includes three territories that occupy almost 3.8 million square kilometres—that's over a third of Canada's landmass. Across this region, those long hours of summer daylight decrease dramatically from August

Beginning on February 8, 1979, residents of Iqaluit, Northwest Territories, were stuck inside for 10 days as a blizzard created zero visibility and temperatures of –40°C.

An aurora borealis display in the Far North.

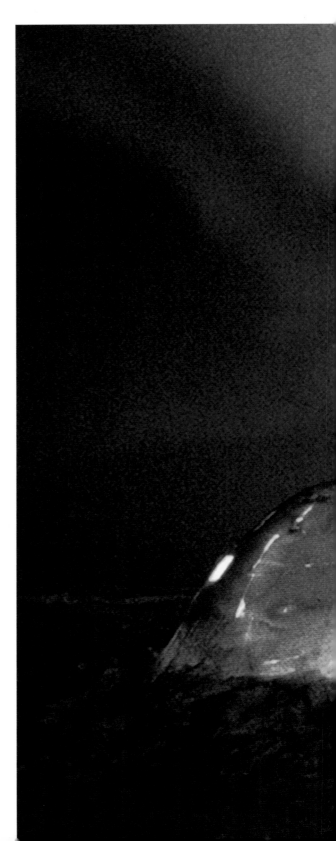

onward, and when the winter solstice finally arrives, the remote communities above the Arctic Circle experience darkness around the clock.

The upside to limited exposure to the sun is that it removes a key element in creating areas of low pressure. Instead, the North gets massive areas of high pressure, as cold, stable air settles to the surface. As we've learned, high pressure has a tendency to remove moisture, and indeed, the air in the Arctic in winter is very dry. The lack of moisture in the atmosphere—added to the fact that lakes, rivers and much of the Arctic Ocean are frozen—means that snowfall is surprisingly minimal. Cold also begets cold—once water and land have frozen, they act like a refrigerator, chilling the air and reflecting what little the sun has to offer.

Look at the average monthly temperatures and snowfall from December to March in these Northern communities.

Average Monthly Temperatures in Some Northern communities

City	Temp.	Snow
Dawson, YK	-21.4°C	20.6 cm
Old Crow, YK	-26.8°C	14.4 cm
Inuvik, NT	-25.8°C	16.8 cm
Yellowknife, NT	-22.8°C	19.3 cm
Cambridge Bay, NU	-31.1°C	6.4 cm
Iqaluit, NU	-25.2°C	21.6 cm

Now compare those with some southern locations, and it's clear that the North is a cold, dry place in the winter.

Wind Chill

On January 28, 1989, the wind chill in Pelly Bay, Northwest Territories, made a -51°C day feel like it was -91°C.

We talk about wind chill all through the winter, and nowhere is it more pronounced than in the North. Yet wind chill isn't a weather phenomenon at all. It's a human concept.

People give off heat, and by doing so, we warm a thin layer of air surrounding us. When we're exposed to wind, especially a strong wind, that layer of warm air is blown away. The body responds by continuing to emit heat in an effort to re-establish that insulating layer, but as long as the strong wind continues, the warm layer of air is quickly removed. We lose heat energy rapidly, and we feel cooler.

But wind does not actually lower the temperature. When the weather report says it's -20°C with a wind chill of -35°C, it means that the thermometer is reading -20°C, but the wind is removing heat from living things at a rate that makes it feel as if the temperature were 15 degrees lower.

If the temperature is only moderately cold — say around -15°C — it takes a strong wind of about 35 km/h to make it feel like -27°C. But when temperatures are very low — as they can be in the North — the wind chill is much more pronounced. An average winter evening in Norman Wells, Northwest Territories, is about -30°C, and a wind of just 15 km/h would have us feeling as if the temperature were really -41°C.

Devon Island, Nunavut.

Wind-Chill Chart

Speed	5°C	0°C	-5°C	-10°C	-15°C	-20°C	-25°C	-30°C	-35°C	-40°C	-45°C	Speed
5 km/h	4	-2	-7	-13	-19	-24	-30	-36	-41	-47	-53	5 km/h
10 km/h	3	-3	-9	-15	-21	-27	-33	-39	-45	-51	-57	10 km/h
15 km/h	2	-4	-11	-17	-23	-29	-35	-41	-48	-54	-60	15 km/h
20 km/h	1	-5	-12	-18	-24	-30	-37	-43	-49	-56	-62	20 km/h
25 km/h	1	-6	-12	-19	-25	-32	-38	-44	-51	-57	-64	25 km/h
30 km/h	0	-6	-13	-20	-26	-33	-39	-46	-52	-59	-65	30 km/h
35 km/h	0	-7	-14	-20	-27	-33	-40	-47	-53	-60	-66	35 km/h
40 km/h	-1	-7	-14	-21	-27	-34	-41	-48	-54	-61	-68	40 km/h
45 km/h	-1	-8	-15	-21	-28	-35	-42	-48	-55	-62	-69	45 km/h
50 km/h	-2	-8	-15	-22	-29	-35	-42	-49	-56	-63	-69	50 km/h
55 km/h	-2	-8	-15	-22	-29	-36	-43	-50	-57	-63	-70	55 km/h
60 km/h	-2	-9	-16	-23	-30	-36	-43	-50	-57	-64	-71	60 km/h
65 km/h	-2	-9	-16	-23	-30	-37	-44	-51	-58	-65	-72	65 km/h
70 km/h	-2	-9	-16	-23	-30	-37	-44	-51	-58	-65	-72	70 km/h
75 km/h	-3	-10	-17	-24	-31	-38	-45	-52	-59	-66	-73	75 km/h
80 km/h	-3	-10	-17	-24	-31	-38	-45	-52	-60	-67	-74	80 km/h

Northern Lights

Auroras are radiant displays of light sporadically observed over the middle and upper latitudes of both hemispheres. They most frequently occur in the polar zone, particularly close to the northern magnetic pole in the Canadian Arctic islands. Here they are known as northern lights or aurora borealis, named after Aurora, the Roman goddess of dawn, and Boreas, the Greek word for north wind. Although unpredictable, the phenomenon most often occurs from September to October and from March to April.

The Sun is constantly emitting solar wind — rarefied hot plasma made up of electrons and ions — in all directions. The Earth's magnetosphere, the region of space that is dominated by its magnetic field, diverts the force of these winds around the planet at a distance of about 70,000 kilometres. Sometimes, however, strong solar winds transfer energy and material into the magnetosphere; these then follow magnetic field lines toward the poles. As these particles move faster, there is an ever-increasing chance that they will collide with atoms and molecules from the Earth's upper atmosphere. When that occurs, energy is lost through the emission of light, producing what is known as an aurora. These can appear as a glow or in an arc, but more typically, they appear as "curtains," with folded parallel rays that line up with the Earth's magnetic field. When different gases interact in the atmosphere, different colours are emitted, but the most common are green, red and blue.

Average Monthly Temperatures in More Southern Cities

City	Temp	Snow
Edmonton, AB	–8.7°C	19.8 cm
Ottawa, ON	–7.2°C	49.5 cm
Fredericton, NB	–6.7°C	58.3 cm
Prince George, BC	–5.7°C	40.8 cm
Quebec City, QC	–9.4°C	65.7 cm
St John's, NL	–3.7°C	65.1 cm

It is a myth that the Inuit have hundreds of words for snow—they have no more than English speakers do. But even though the North gets less snowfall than many other places in Canada, it's part of life in northern communities for much longer. It generally arrives in late September and stays until the end of April—so there is often snow on the ground and ice in the lakes from just after Labour Day until almost May 24.

The most difficult and dangerous aspect of winter in the North is the combination of cold and wind. Eureka, Nunavut, recorded the coldest month on record in Canada: the average temperature was –47.9°C in February 1979. Quaqtaq, on the shores of the Hudson Strait in northern Quebec, recorded Canada's strongest winds in November 1931: 201 km/h.

The reason it can be so windy has to do with what's happening in the upper atmosphere. Cold, high-pressure centres at higher altitudes oscillate through the North, a bit like the action of a wave pool. As they move through an area, they cause changes in pressure and temperature, leaving strong winds in their wake.

Strong winds, blowing snow and severe cold can disorient and kill people in short order. That's why the Inuit learned to

The Inuit have adapted to their environment and learned to use snow to make igloos, a very efficient shelter against the harsh winter weather.

build appropriate shelters. While nature provided them with little wood or stone, the weather provided a unique type of building material. Snow in the Arctic is compressed and dried by wind and cold air to a consistency that will hold its form when shaped. It's solid enough to cut the wind, yet it contains enough air pockets to offer insulation. That makes the igloo one of the most efficient types of shelter. In an emergency, digging a hole in the snow and crawling in will save your life.

The Prairies

BEFORE WE LOOK at winter in Alberta, Saskatchewan and Manitoba, we need to understand that what we call the Prairies—from the foothills of the Rocky Mountains to the Ontario border, and from the 60th parallel to the U.S. border—is really two different topographic regions. Imagine a line running from Lesser Slave Lake in Alberta to Lake Winnipegosis in Manitoba. The area that lies south of this line is a true prairie, with rolling grasslands interspersed with rivers and lakes. The area north of the line, however, is forested and covered with thousands of lakes. All of the area is relatively flat, however, and that's important to the climate of the Prairies.

The most dominant winter feature in this region is arctic high pressure, which circulates cold, dry air south and eastward. The polar jet stream defines the southern limit of this cold air, and at the same time steers moisture and warmer air in from the Pacific. When the arctic high pressure is strong, typical winter weather prevails, but when the high shows signs of weakening, it makes for some interesting scenarios.

When warm, humid air from the coast crosses the Rocky Mountains into Alberta, most of its moisture gets exhausted, and a subtle temperature difference exists between the Pacific air and the colder air that lies over the Prairies. This creates an area of low

In Winnipeg, a heavy snowfall is twice as likely to occur on a Monday than on any other day of the week.

A snowstorm rips across the Prairies.

Cold air
behind storm

Alberta
Clipper

Alberta Clippers

Snow in the Great Lakes region often arrives courtesy of Alberta clippers. These fast-moving storms originate with a low-pressure region that develops just east of the Rocky Mountains. The cold, stable air mass in the Arctic then propels the storms swiftly eastward. Alberta clippers usually leave behind only a light snowfall, but their high winds can bring biting cold.

About one septillion (1 followed by 24 zeros) snowflakes fall on Canada each year.

pressure and can develop into storms. When the jet stream winds are strong enough and moving in the right location, these storms get spirited eastward at high speed, giving rise to their name: Alberta clippers. They can travel from their breeding ground to the Great Lakes basin in less than a day and a half. When they reach farther east, they have a tendency to become even more potent storms.

The clipper storms that race across the Prairies are common in winter—often they'll come in waves, one after another, with just a day or two between. As they cross the western landscape, each will put down five to 10 centimetres of snow, and the ensuing northerly winds can create near blizzard conditions. As the low rushes to the east, it also draws colder air from the Arctic, and from higher altitudes, creating very strong winds.

The clipper pattern is usually broken when the arctic high-pressure ridge shifts

farther south, or when an unusually strong storm reaches the coast of British Columbia. Powerful Pacific storms force increasingly milder air over the Rockies, and that brings a change in Prairie weather, especially in Alberta.

It's an old joke that winter temperatures in the Prairies can be bitter, "but it's a dry cold." For the most part, that's true. Because of the relative lack of large bodies of water in this region—especially in Alberta and Saskatchewan—little moisture is available. In winter, the air is so cold that almost all of the moisture has been condensed into ice crystals that offer very little in the way of relative humidity. Compare these average snowfall rates:

Average Snowfall Rates in Some Prairie Cities, Versus More Western and Eastern Cities

City	Snow
Red Deer, AB	116 cm
Saskatoon, SK	97 cm
Brandon, MB	112 cm
Williams Lake, BC	192 cm
Hamilton, ON	161 cm
Moncton, NB	349 cm

The right amount of snowfall is critical to the many Prairie farmers whose livelihood is tied to the land and its weather. Snow cover in winter protects the soil from the eroding power of the wind. Melting snow in spring provides a layer of moisture for plantings. If there isn't enough snow, farmers lose both their precious moisture and their precious soil. But if there's too much snow, an uneven thaw presents the possibility of flooding when spring arrives.

Chinooks

A warming wind that blows out of the mountains is called an adiabatic wind. In western Canada, when a strong current of warm, humid air flows eastward from the Pacific Ocean, it eventually meets the Rocky Mountains, where it is forced upward. As it rises, moisture is condensed, and it falls as snow or rain on the western side of the mountains.

Now this dry but warm air flows over the crest of the mountains and begins to push toward the surface on the eastern side of the Rockies. As the air is forced toward the surface, it compresses and warms further. As the heating process accelerates, it is not uncommon for the surface temperature to rise 30 degrees in just a few hours.

In Calgary, chinooks occur several times every winter and drive the temperature from –10°C in the morning to 20°C by two in the afternoon. That's why we sometimes see Albertans wearing shorts in January. Unfortunately, the one guarantee with a chinook is that it lasts a very short time — perhaps only a day or so.

The word chinook does not come from an aboriginal word that means "snow eater," as some people believe. The Chinook were a people who inhabited what is presently Oregon and Washington. When these warm winds blew from the mountains and brought relief from frigid weather, the inhabitants of the plains named the winds after the people who they believed had sent them as a gift.

Wind

Dissipating clouds

4,200 m
–15°C

914 m
18°C

Chinook Wind
Adiabatic heating
–14°C per 304 m

Clear and dry

Chinooks

Southern Albertans welcome the arrival of chinooks, which can raise winter temperatures by as much as 30 degrees in just a few hours. A chinook occurs due to a process called adiabatic warming. When moist air is driven against the windward side of a mountain, it rises and condenses, often causing heavy precipitation. The drier air is then forced down the leeward slope, where it compresses and warms further, causing high winds and a spike in temperature.

A fabulous sunset ends the day after a chinook blows through Alberta just outside Calgary.

This photo of an early-20th-century ice storm in Elora, Ontario, shows the destruction that freezing rain can cause.

and slow-moving Colorado low siphons the warm air toward the Great Lakes, the result is often potent snowstorms with strong winds because of the great pressure and temperature changes involved. The storms leave behind up to 20 centimetres of snow and precipitate a rapid drop in temperature.

Most Canadian winters are cold enough that much of the surface of the Great Lakes will ice over. Only the deepest parts of Lake Superior and Lake Ontario stay ice-free during an average winter. This open water actually works as a heat reservoir, creating a unique microclimate in Ontario. Ice doesn't usually form in the lakes until mid- to late winter. Until then, the lakes supply enough heat to keep the air over the water a few degrees warmer than the air farther inland.

This phenomenon is a mixed blessing, however, as it also produces lake-effect snow.

Because so many people live in this part of our country, transportation is critical. Nothing can bring the travelling public to a halt faster than that other hallmark of a central Canadian winter: frozen precipitation. This can fall either as freezing rain or ice pellets, though the former form creates far more havoc on the roads.

Freezing rain can only occur when the temperature at the surface is below zero, while the air above is warmer. Liquid rain falls into a shallow layer of very cold (at least –1°C) air, but because it passes through so quickly, it does not freeze. The droplets become supercooled—that is, they are chilled to a sub-zero temperature without becoming solid. As soon as the liquid rain lands on a surface, such as the road or your car, it instantly freezes. The result is a smooth, clear layer of ice. That's where the danger lies: we see and feel the rain, but it isn't until we drive or walk on a surface that we realize it has become ice.

In the case of ice pellets, the layer of cold air is much thicker, and the rain actually does freeze solid. The surface temperature, meanwhile, can be above or below zero. Think of it this way: you can shovel ice pellets, but you can't shovel freezing rain.

By the time mid-January arrives in Ontario and Quebec, the days are becoming noticeably longer, and the third week of the month often brings a mysterious event known as the January thaw. The January thaw is an interruption in the usual weather patterns, and it can temporarily chase away the winter blues, but it's

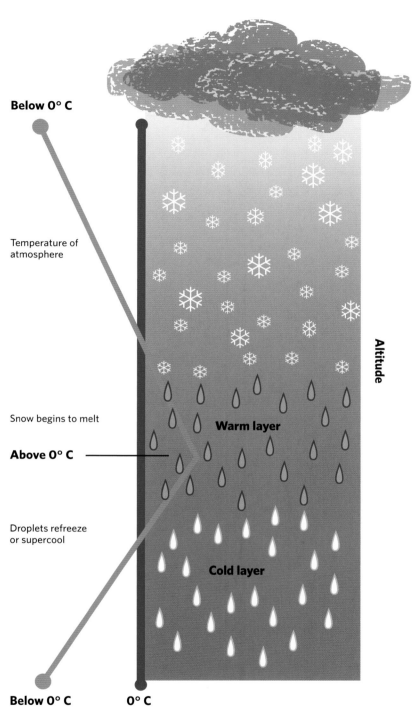

Freezing Rain

When the temperature at the ground is below zero, but the air above is warmer, conditions are ripe for one of the most dangerous types of weather. Solid precipitation melts as it falls through a warm layer. The liquid then continues through a shallow layer of very cold air, where the droplets are chilled to a sub-zero temperature without becoming solid. When these supercooled droplets hit the surface, they instantly freeze, creating hazardous road and sidewalk conditions.

Ice skating on the Rideau Canal in Ottawa.

short-lived: until the sun is higher in the sky, the stronger arctic high pressure will eventually return to dominate.

One of the snowiest places in Ontario is the town of Wiarton, home of the groundhog Wiarton Willie, who gets his day in the sun every February 2. Perhaps Groundhog Day should be a national holiday. Think about it: it's not based on religion, politics or history. It's just a celebration for everyone who loves winter— and for everyone who despises it.

It's an offering of hope to all who gather to wait and hear what a rodent will tell us.

Whether or not the famous groundhog sees his shadow, spring doesn't arrive for another six weeks, of course. But the story does have at least a little meteorological truth to it. A clear day in early February—when Willie would see his shadow—indicates that arctic high pressure is the dominant weather pattern, which means the cold is here to stay for a while. An overcast day with no shadows indicates low pressure and a warmer, more southerly weather pattern.

Lake-Effect Snow

Unlike other kinds of precipitation, lake-effect snow isn't associated with low-pressure centres. Instead, it comes to us when there's a strong prevailing wind over open water.

Aside from the open water, we need two other ingredients for lake-effect snow. The first is a large temperature difference between the water and the air mass that moves over

it; ideally the air should be at least 10 degrees colder than the water. The air also needs to be fast and unobstructed as it moves over the lake.

The dry, cold air absorbs moisture from the water, which causes clouds to form. When they become saturated with water droplets, which freeze to become flakes, they release the moisture as snow. A

lot of it. The lake effect can present nearly continuous, accumulating snowfalls.

Lake-effect snow can be extremely dangerous for drivers, because the snows are wind-driven, and the winds have a tendency to shift directions. A clear day that begins with a pleasant drive to work can quickly change to a precarious commute home in a raging blizzard.

Built along the waterfront, the city of Toronto is subject to lake-effect snow.

The Atlantic Region

HERE'S A SAYING that perfectly describes the weather conditions year-round in the Atlantic provinces: "If you don't like what you see out the front window, then look out the back door."

Growing up in Halifax, I saw just as many green Christmases as white ones. I remember wishing for a break from the rain and fog. I also remember clearing massive amounts of snow from our walkways.

Moncton, New Brunswick, averages more snow than any other major city in Canada.

The east is in a unique position to receive both modified arctic air and an abundance of moist, warm Atlantic air. When combined, the two create the biggest winter storms in Canada. Compare these average winter temperatures and snowfalls for a few Atlantic communities with those of towns farther west:

Average Winter Temperatures and Snowfalls

City	Temp.	Snow
Halifax, NS	–3.9°C	230 cm
Sydney, NS	–4.2°C	298 cm
Saint John, NB	–5.6°C	256 cm
Moncton, NB	–5.7°C	295 cm
Charlottetown, PE	–5.7°C	311 cm
Gander, NL	–5.9°C	443 cm
St John's, NL	–3.7°C	322 cm
Montreal, QC	–6.8°C	217 cm
Toronto, ON	–3.7°C	115 cm
Calgary, AB	–6.1°C	126 cm
Kelowna, BC	–1.4°C	102 cm
Winnipeg, MB	–12.9°C	110 cm
Edmonton, AB	–8.7°C	123 cm

Although this scene in Hunter River, Prince Edward Island, is picturesque, the weather in the Atlantic region is notoriously unpredictable.

This 2007 snowstorm in St. John's, Newfoundland and Labrador, shows how completely nature can bring a modern society to a halt.

Freezing drizzle contributed to the crash of a DC-8 in Gander, Newfoundland, on December 12, 1985, that killed 248 U.S. soldiers.

Very few communities in Canada get as much snow as those in the Atlantic provinces. The reason is the nearly constant meeting of cold dry air and moisture-laden mild air from the ocean. Remember, even the most inland areas in these provinces—with the exception of Labrador, which has a nearly Arctic climate—are no more than 200 kilometres from the sea.

The Gulf Stream keeps the ocean temperature at a reasonable 4°C to 8°C off the coast of Nova Scotia. This warm water is the engine that helps intensify storms along the East Coast. This push of arctic air into the relatively warm Atlantic creates some potent storms. They have many names—nor'easters, sou'westers, weather bombs—and all have a common root cause.

With all this extreme weather, why is there often little or no snow on the ground in places like Halifax and St. John's? This fact takes us back to the saying about the weather being different out the front window and the back door. The warm Atlantic air does a very good job of melting the snow it just deposited—or of following up a healthy 20-centimetre snowstorm with several hours of steady rain. Often, it happens like this: a powerful nor'easter with nearly hurricane-strength winds and 20 or more centimetres of snow develops. This storm then draws the mild air from the sea over the land. Now all the beautiful white snow turns slushy as steady rains fall and thick fog envelops the area.

The warm ocean also serves to moderate the temperature along the coast. The daytime averages in Halifax, St. John's, Sydney and Summerside hover just below the freezing mark during January and February.

When strong storms move into Atlantic Canada from along the coast, they often become much more unpredictable in regard to what precipitation they will bring. If there is a thick enough layer of cold air at the surface, there will be snow. When the layer is shallower, the chance of freezing rain increases. If the surface air is above freezing, the weather may begin as snow but will quickly transition to rain. The low pressure centres that arrive in our eastern provinces from along the Atlantic coast tend to be much milder because the temperature gradient is weaker, often due to a weakening of the continental arctic air over Quebec.

A major 2004 winter storm in Halifax, Nova Scotia, nearly obscures the Old Town Clock from view.

Weather Bombs

In an average storm, the pressure usually drops only about 10 or 12 millibars in 24 hours. When the barometric pressure drops 24 millibars in 24 hours, the resulting storm is called a weather bomb, a fitting description for this incredible and intense change.

The reason for this pressure drop is the influx of very cold air over the comparatively warm Atlantic Ocean. This push of cold air is further assisted when the low pressure begins to circulate cold air from the Labrador Current into its deepening core.

Often the air over the Atlantic will be in the order of 10°C, while the incoming arctic air will be 20 degrees colder. This huge temperature difference is what forces the storm to implode. Because there is a great temperature change, there is also a great pressure change, which results in hurricane-force winds and lots of moisture for snow.

Weather Watching

E VERYONE WANTS TO know what the weather is going to be. Whether they're planning a vacation, spending a day at the golf course, looking forward to a weekend of gardening or just heading to work or school, millions of Canadians check the weather forecast every day on television and radio, in our newspapers, on the Internet, even on our cell phones and PDAs. Let's look at how the information about our weather is gathered, analyzed and presented to you.

I work for The Weather Network, the largest private-sector provider and distributor of weather information in Canada, and one of the largest in the world. Before we can do our jobs, however, we need weather data, and gathering that raw material is the job of Environment Canada. Trained weather observers from coast to coast record what is happening, usually at the beginning of every hour. Some of these observers are meteorologists at official weather stations, airports and seaports. There are also trained civilian observers scattered across the country. Aviators and mariners report weather conditions regularly too. There's also a network of automated weather stations, both on land and at sea, that constantly relay information to meteorologists at Environment Canada.

To learn what's going on in the atmosphere, we use specialized tools to measure several conditions.

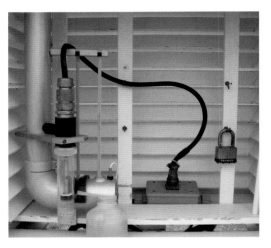

Left: Stevenson screen. Right: Interior of Stevenson screen.

Official temperature is measured at the "thermometer screen," about 1.8 m above the ground.

Temperature

To measure temperature, we use the same instrument you have outside your kitchen window: a thermometer. The location of the weather station will dictate the type of thermometer used—the extreme cold of the Arctic, for example, makes mercury-based thermometers useless, so alcohol-based thermometers are employed there. Weather stations have a small shelter called a Stevenson screen, which houses many instruments, including the thermometer. If it were placed in direct sunlight, it would register a higher-than-true temperature—that's why your home thermometer should be in a location that is always in shade. At many weather stations, there is also a special thermometer designed to record only the highest and lowest temperatures in a 24-hour period. To gain an understanding of future weather conditions, it's important to track trends—rising temperatures often announce the arrival of low-pressure systems, for example.

Wind Speed and Direction

To measure wind, we use an anemometer. You've likely seen this little device—it's a spinner with three little cups on it. The cups capture the wind and cause the device to rotate; the rate of that rotation is the wind speed. A vane is attached to show us the direction of the wind. An anemometer should be positioned away from ground obstructions that may alter the speed and direction of the wind—it is typically placed on a mast about 10 metres above the ground. The wind doesn't always move at the same speed or direction, and when the observations are taken hourly, we may miss some of how the wind has behaved since the previous measurement. Most modern anemometers take this into account and give an average speed and direction that is computed every few minutes. The speed of wind is measured in either metres per second or knots (nautical miles per hour). The direction the wind is blowing from is expressed in degrees in relation

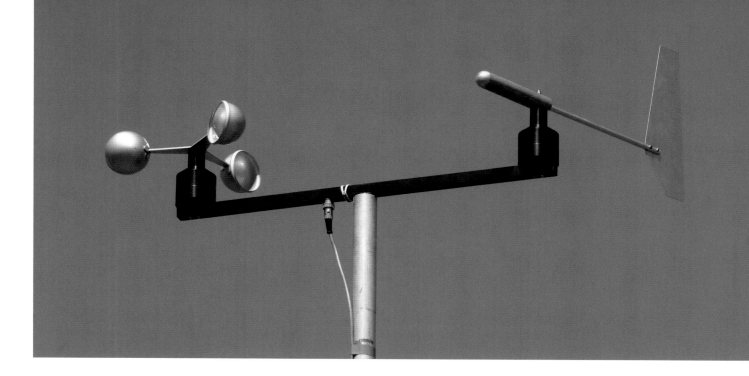

Above: Anemometer

to the magnetic north pole—a wind from due west is given as 270 degrees. When measured over time, the speed and direction of wind give us information about the location of weather systems and their stage of development.

Pressure

A barometer measures atmospheric pressure. The variety many of us have in our homes is known as an aneroid barometer, and it can often indicate—at least in a broad sense—whether we can expect fair or rainy conditions. In the 1640s, Evangelista Torricelli, a student of Galileo, created the first barometer by placing a sealed tube filled with mercury into a basin that also contained mercury. As the atmospheric pressure rose and fell, the mercury in the tube moved accordingly. The more accurate barometers we use today are corrected for subtle changes in temperate and elevation, but they

still work on the principle that Torricelli recognized nearly 400 years ago. A barometer's tube is calibrated to average sea-level pressure, which is 1,013 millibars, or 76 centimetres—that's how far the mercury in the tube rises under normal conditions. By tracking the movement hour by hour, we are able to determine whether the pressure is falling (indicating a low-pressure system and, potentially, rain) or rising (indicating high pressure and fair weather). The lines on a weather map that connect equal barometric pressure readings are called isobars, and they offer us insight into wind direction and speed, as well as longer-term patterns over large areas.

Nor'easter is the name given to a storm formed off the eastern North American coast by the interaction of cold arctic air and warm, moist air over the Gulf Stream.

Q Which Canadian community enjoys the calmest weather?

A In Kelowna, British Columbia, 39 percent of all weather observations feature a wind reading of calm.

Visibility

Visibility is the distance we can see toward the horizon. It was traditionally measured by using the naked eye and a series of benchmarks, but a far more accurate method employs a laser that has been calibrated to simulate human vision. At land-based weather stations, visibility is measured in units of 100 metres to a distance of five kilometres, then in one-kilometre units up to 30 kilometres, and five-kilometre units beyond that. (For visibility of less than 100 metres, increments of 10 metres are used.) Measuring visibility—which is vitally important for aviators and mariners—gives us an idea of how cluttered the atmosphere is with water vapour, fog, smoke and even pollution. The lower the visibility, the more moisture is in the atmosphere and the greater the likelihood of precipitation.

Q Which Canadian city receives the most sun each winter?

A Despite its reputation for biting cold, Winnipeg receives an average of 358 hours of mainly clear, sunny skies from December to the end of February. That's more sunshine than any other city in Canada gets.

Precipitation

The rain gauge is probably the simplest weather tool in the world. It's simply a container that catches rain and snow and funnels it through a tiny aperture

The rain gauge is one of the world's simplest weather tools.

to a measuring vessel. Rain is measured in millimetres, while snow is measured according to its liquid equivalent—10 millimetres of snow equals about one millimetre of rain. The container must be at least 12 centimetres tall and its base must be elevated at least 12 centimetres so no water hitting the ground will splash into the gauge. A rain gauge is also placed away from buildings or trees, which can lead to inaccurate readings. At least once a day—hourly when there is steady rain or snow—the observer takes a measurement of how much precipitation has fallen. Today, much of this process has been computerized, and the observer can check the reading any time without having to go out into the wet.

The Campbell-Stockes recorder measures the amount of sunshine in a day.

Humidity

Humidity is measured with an instrument called a hygrometer, which detects the amount of moisture in the air, and it can be expressed in a couple of different ways. Absolute humidity is the mass of water in a given volume of air. A more useful measure for weather forecasts, however, is relative humidity, which indicates how close the air is to being saturated. Relative humidity is expressed as a percentage: 50 percent means the air contains half of the moisture it is capable of holding at the current temperature. That's about the level at which we're most comfortable. Because high humidity in summer can make it feel hotter than it really is, Canadian meteorologists invented the

"humidex" in 1965. This measurement combines relative humidity and temperature to indicate the level of comfort most people will feel on a summer day. (A humidex of less than 29 is pleasant, while a reading in the 30s is sticky, and a day in the 40s can be downright dangerous.)

Sun Exposure

Weather stations also measure the amount of sun we get every day. In addition to timing sunrise and sundown, meteorologists use a Campbell-Stokes recorder, which consists of a glass sphere mounted in a frame so that it focuses sunlight onto a card marked with lines indicating the time of day. As the sun moves across the sky, the focused beam burns or scorches a line across the card—the longer the line, the more

The term smog was coined in 1905 as a combination of smoke and fog.

Toronto residents enjoyed a daytime high of 11°C on January 27, 2002.

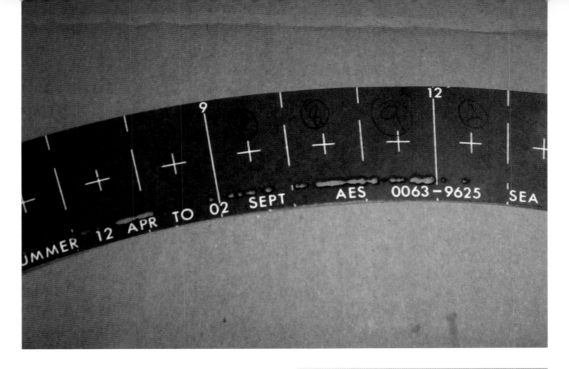

Close-up view of a summer sunshine card for the Campbell-Stokes recorder. The amount of sunshine is recorded in tenths of an hour.

sunshine was received during the day. During cloudy periods, when it's not bright enough to burn the paper, an instrument called a thermopile can measure the intensity of solar radiation at the surface in watts per square metre.

Level of Cloud Cover

0 oktas	Clear skies
1 oktas	Almost clear skies, just the odd cloud
2 oktas	Mostly clear skies, only a quarter of the sky covered by cloud
3 oktas	Partly cloudy, just over half the sky is cloudless
4 oktas	Partly cloudy, half of the sky covered by cloud
5 oktas	More than half the sky covered by cloud
6 oktas	Mostly cloudy, only a quarter of the sky showing
7 oktas	Almost overcast, just a small amount of sky showing
8 oktas	Overcast, no sky showing

Clouds

Weather observers also note the clouds by type and altitude. This is a rather simple observation that involves identifying the cloud type and noting what percentage of the sky is covered. Meteorologists divide the sky into eight sections called oktas, and cloud cover is expressed as a number between 0 (clear sky) and 8 (completely overcast). In the past, the altitude of clouds was determined by releasing a balloon and timing its ascent. Today, we use lasers to gather the same information on the altitude of various cloud layers. The type and elevation of clouds, as well as the direction they are moving, can go a long way in helping us understand what the weather is going to do. For instance, gathering clouds signify more moisture in the atmosphere and a greater chance of precipitation, while dissipating clouds are indicative of drier air and fair weather.

This cloud cover over a drilling station
on the Prairies rates about 2 oktas.

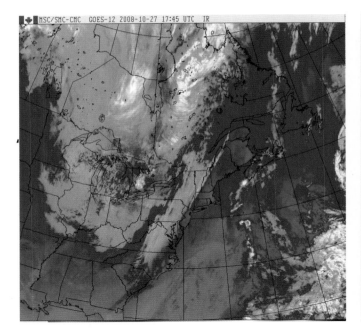

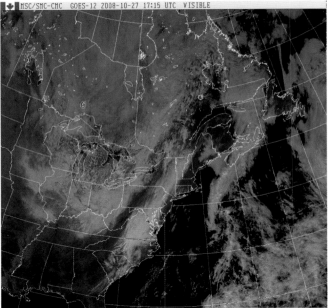

THESE TOOLS WE'VE described are employed at weather stations across Canada and around the world. For almost 50 years, however, we've also been able to get a perspective on our weather from space. The world's first successful weather satellite, TIROS-1, was launched on April 1, 1960, and today, is a constellation of weather satellites orbits Earth. In addition to monitoring cloud cover, they measure snow and ice cover, atmospheric moisture, sea-surface temperature, even pollution and volcanic ash in the atmosphere. Early probes were stationed about 800 kilometres above us, but modern weather satellites are now about 35,000 kilometres above the equator. These satellites are in geostationary orbit, meaning that they move at a speed that matches the rotation of the planet, so they're always above the same point on Earth. At The Weather Network, we use images from Geostationary Operational Environmental Satellites (GOES) East and GOES West, two satellites that look down

Images from the GOES East satellite. Left: infrared; right: visible.

on eastern and western North America, respectively.

We also use radar technology to measure what is falling from the sky. Weather radar works by emitting radio waves—when these waves strike something, they bounce back toward the ground, allowing scientists to determine the speed and direction of the object that caused the reflection. In this case, the object is precipitation. The sensitivity of the radar's beam can be adjusted so analysts can determine the type of precipitation and the intensity with which it is falling, enabling us to estimate the amount of rain or snow that will accumulate on the ground. By monitoring the direction and intensity of precipitation, we can forecast how storms are evolving and what potential hazards they may bring. Radar is the best tool meteorologists have for getting real-time

At −18°C, a snow-mobiler travelling 37 km/h would feel a wind chill of −42°C.

information on active weather—it's updated by the minute and allows us to create a three-dimensional image of what's happening, as well as what's likely to happen in the short term. In Canada, there are both private and government radar sites dedicated to tracking weather. The Environment Canada network covers our country from Pacific to Atlantic, and very few populated areas are without coverage.

At The Weather Network, we have access to all the information gathered by Environment Canada, as well as satellites and radar sites. Once we've gathered all of this information, we combine it to create a weather map that shows the current state of the atmosphere. In one hour, we'll create another weather map with all of the updated information. This is the first step in putting together a forecast.

Although humans have always tried to predict the weather, scientific forecasting has a relatively short history. In the early 20th century, scientists believed they

Above: Meteorologists must evaluate and combine many different sources of weather information in order to develop the most accurate forecast possible.

Below: A radar weather map.

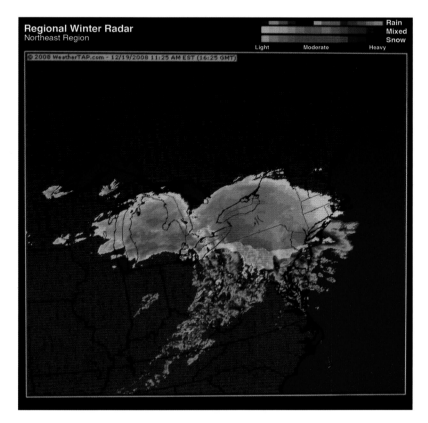

Cameraman and reporter filming on location during the first big snowstorm of the winter season in 2008.

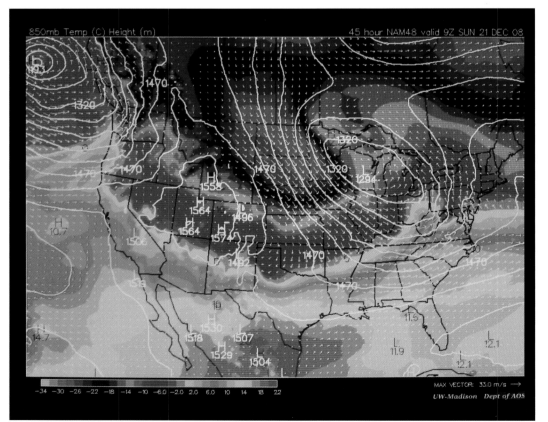

Weather map showing projected weather patterns.

could use mathematical formulas to determine how the weather would behave. The trouble was that the equations were so complex that by the time the scientists had completed the math, the weather they wanted to forecast was already occurring. By 1922, British scientist Lewis Fry Richardson was able to create a crude formula that combined weather observations from several areas and the historical data from those weather sites and produced a reasonably accurate prediction for the next day. This was the beginning of what's called numerical forecasting—using math and history to determine what might happen in the future.

The development of the computer made the process of gathering and compiling information much easier. In the 1960s, it became possible to make models of the atmosphere's future behaviour by using data about its present condition and factoring in historical data. Now, with years of experience, far more powerful computers and vastly more detailed data available, we can create far more detailed and accurate projections—not only for tomorrow but for the days and even weeks ahead. Of course, there are still plenty of surprises, but short-term forecasts are right about 90 percent of the time. We can forecast about five days in advance with around 50 percent accuracy.

There is one thing that computers can't do and that is understand the nuances of local weather. I recall a time when I was the weatherman at a television station in Kingston, Ontario. Before presenting the weather on the evening news, I would have a conference call with an

Toronto and Tokyo average almost the same amount of sunshine each year — 2,038 and 2,021 hours, respectively.

The author in the studio stands against a green screen, upon which the map image will be superimposed.

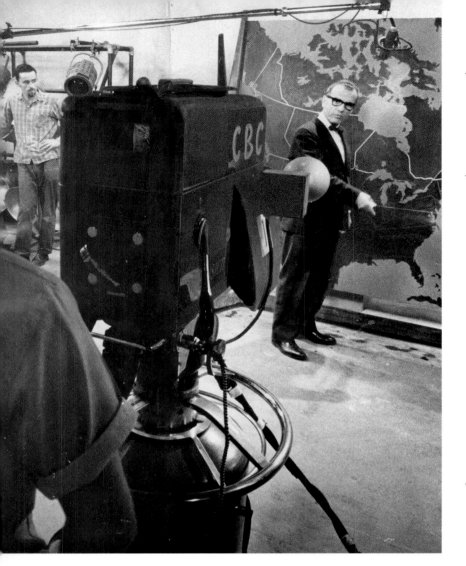

Percy Saltzman was the first Canadian to appear on English-language television and the CBC's first meteorologist.

On average, it's foggy at Old Glory Mountain, British Columbia, 226 days each year.

Environment Canada meteorologist to discuss the forecast for eastern Ontario. On this particular early-winter day, the forecast was for strong southwesterly winds, and the meteorologist, being new to the region, didn't pick up on the fact that this can produce what every Kingston resident knows as lake-effect snow. I relied on my experience and warned the city to expect flurries—and that's exactly what happened.

Often the forecast you see on The Weather Network differs from the one issued by Environment Canada. That's because we create our own forecasts using not only the data from our own government but also that of the National Weather Service in the United States and the Met Office in the UK. Our staff of meteorologists may, for example, feel that the British computer model offers a truer picture of what will happen. Other times, they may believe that a blend of the Canadian and American models will give a better forecast. Once the decision is made, weather maps and a text version of the information are prepared, and we're ready to make it public. The Weather Network has developed groundbreaking technology to distribute local forecasts as part of a national television broadcast. From our data distribution point in Toronto, we take the information for each city and relay it by satellite to computers across Canada. As a result, if you watch cable television anywhere in the country, you'll get a forecast specific to your area.

The actual presentation of a weather forecast is an exercise in economy. We take all that data and distil it into a few words and symbols: "Expect a sunny day with a high of 18°C tomorrow." The process begins with a series of meetings between the meteorologists, presenters, writers, researchers and television producers. This team determines the most interesting and significant weather of the past 24 to 48 hours and decides which maps, video footage and other images we can use to illustrate this part of the story. We then determine what will be the most important weather story in the next 24 to 48 hours. Covering big storms is critical

The combined image of the weather map and the meteorologist, as seen by television viewers.

for public safety—a blizzard in a densely populated area can potentially affect the lives of millions of people. If a big storm is brewing, our producers and researchers stay in touch with local officials and emergency services, as well as camera crews and our network of local weather watchers. Finally, our team looks ahead to the longer-term forecast. How long will this heat wave, storm, drought or cold snap last?

Once we've assembled these elements, we go to air. A lot has changed since Percy Saltzman first presented a weather forecast on CBC television in 1952. In the beginning, Saltzman, who made more than 9,000 weather broadcasts in his 30-year career, had only a chalkboard and his unique storytelling style. Today, we have the magic of computer animation and imaging. Although you see me standing in front of a weather map on your television screen, I actually work in a studio that is painted either green or blue. There are no maps and no backgrounds, just an empty room with cameras and television monitors. A computer combines everything in layers, and the monitors in the studio show me where I am in relation to the background images, so I see the same finished product you see at home.

The process of gathering information, analyzing data, preparing forecasts and finding the weather stories that are most important goes on 24 hours a day, every day of the year. It goes on because the story—like the weather itself—never stops unfolding.

5

Canadian Weather Disasters

AS OUR GLOBAL climate continues to change, so, too, do the local weather patterns that affect our country and its people. The weather plays out each season with its usual array of storms, drought, floods, heat waves and cold spells. But every so often, the routine is broken by once-in-a-lifetime meteorological events that we will later tell our grandchildren about.

The recent weather events I've assembled here serve to highlight both the power of nature and the resilience of Canadians. This is not a top 10 list but rather a look at the enormous and unforgiving strength of our environment— and how insignificant we can become in the face of it. Every time we endure one of these great trials, it further fortifies our respect for the power of nature and defines our role in the evolution of our planet.

In the coming decades, I am sure we will see more dramatic occurrences as the great aerial ocean changes, churns and issues forth weather events that will alter our ideas of snow, rain, heat, cold, wind and more.

An aerial view of an over-flooded dam on the Chicoutimi river July 21, 1996; shows a house resisting a major flood in downtown Chicoutimi.

Okanagan Mountain Park Fire

THE SUMMER OF 2003 in the southern interior of British Columbia was one of the driest in a century. Between June 22 and August 15, only 2.2 millimetres of rain had fallen in Kelowna—hardly enough to notice in a rain gauge. It was also relentlessly hot: the mercury routinely crept above 30°C from late June onward.

The forest service was keenly aware of how extreme the weather had become by mid-August, as were the citizens of the beautiful Okanagan Valley. Crews had been busy battling small blazes for more than a month, and firefighters remained on high alert as the persistent high pressure that was centred over the province showed no signs of weakening. The valley had become a tinderbox, and all it would take was a spark to set it ablaze.

Just before 2 a.m. on Saturday, August 16, a dry lightning strike occurred on Rattlesnake Island, south of Kelowna in Okanagan Mountain Provincial Park. The fire was reported within 30 minutes, but by the time aerial attacks began at sunrise, it had already grown to 15 hectares. Firefighters cut away trees to create a barrier free from fuel and waded into the blaze to turn over the soil, while aircraft dropped water and flame retardant. By afternoon, however, the winds began to increase to about 35 km/h, and the heat from the fire was also creating its own wind. The inferno grew to over 200 hectares by

nightfall, and although the fire was still several kilometres from the nearest homes, the threat of it spreading into the city was real. Over the next few days, the City of Kelowna and provincial forest firefighters began preparing a firebreak that would be nearly 17 kilometres long and 50 metres wide. They hoped this clearing would protect homes on the southeast perimeter of the city.

By August 20, the fire covered nearly 9,000 hectares and had consumed several communications towers. To this point, the winds had generally been blowing from the north, but then the weather changed: the gusts started to blow from the south and west, pushing the massive fire northeast toward Kelowna. Evacuation orders were issued for subdivisions on the southwest side of the city, and crews tried removing trees and spraying water and flame retardant on the structures. By August 22, 17 homes had been saved from the fire, but 21 had been destroyed. Then, just before 5 p.m. that day, the fire exploded in intensity as winds and downdrafts of over 75 km/h sent the blaze into a wild frenzy. Burning debris was falling up to eight kilometres ahead of the main fire, and these in turn started smaller fires.

The winds didn't ease until just after 3 a.m. the next day. Finally, the firestorm lost some of its fury. When the sun rose that smoky Saturday, a full week after

On August 15, 1958, extreme heat from a forest fire created a tornado in British Columbia, killing two firefighters.

the fire had begun, close to 250 homes had burned to the ground. It was a miracle that no one died. The fires would not be extinguished until the end of September. Before it was finally doused, the Okanagan Mountain Park fire had destroyed an area of 25,000 hectares.

On August 16, 2003, the Okanagan Mountain Park fire consumed much of the standing forest.

Ontario-Quebec Ice Storm

The 1998 ice storm ended on January 10, after more than 80 hours of freezing rain and drizzle. It brought down 120,000 km of power lines and phone cables and was responsible for more than two dozen deaths.

NEW YEAR'S DAY 1998 was cold in eastern Ontario and western Quebec. Snow had fallen during the last few days of December, and on Friday, January 2, there were 30 centimetres on the ground in Montreal and 32 centimetres in Ottawa. It seemed that it would be a typical January. But by Saturday, the January thaw had come early. The temperature in both cities was more than 10 degrees above the usual −5°C daytime high. It was a drizzly and mild weekend, and by Sunday night, the snow had melted away to just a few centimetres.

Then the cold returned with a vengeance. A powerful area of high-pressure lay in the Atlantic near Bermuda and another farther north over Labrador, and the circulation pattern began funnelling this cold air southwest. As it rushed into southern Quebec and eastern Ontario, the temperature dropped from 7°C on Sunday afternoon to nearly −10°C that night.

Monday morning began with light freezing drizzle, leaving people to scrape their car windows and proceed with caution on roads and sidewalks. Crews sanded and salted to keep the roads and sidewalks safe. By day's end, more than 17 millimetres of ice coated an area that extended from Kingston north to Ottawa, and then eastward through Montreal, southern Quebec and northern New Brunswick. The ice cover went as far south as northern New York, Vermont,

New Hampshire and Maine. In all, thousands of square kilometres of eastern North America were under a layer of ice that continued to accumulate.

By midweek, not only were the power lines breaking under the strain of ice, but the poles and transmission towers were collapsing, leaving some four million people without power. The eerie swishing sound of the freezing rain was punctuated with the explosions of trees as they snapped and fell under the weight of tonnes of ice. Overpasses were closed because of fears they might collapse. Without electricity, people began to heat their homes with fireplaces, generators and barbecues, and the threat of carbon monoxide poisoning grew. Twenty-five people died. Away from the cities, barns collapsed, killing the livestock that hadn't already frozen to death.

I worked at The Weather Network during this storm, and our studios were located in Montreal. I remember the drive to work from Kingston on Friday, January 9, just as the storm was easing. It looked as though someone had taken a giant scythe and sliced off the treetops. The electricity was off more than it was on in my Montreal hotel, and streets were closed because so much ice was falling from the buildings.

By Monday, January 12, the weather had completely changed. The freezing rain was over, and the low pressure that drew in the moisture had moved across the Atlantic provinces, where it dropped an average of 30 centimetres of snow. Cold air settled into eastern Ontario and Quebec, the mercury would not climb above zero again until the first day of February. Only snow would fall on this icy landscape for the next few weeks as people began to survey the damage. The cost would eventually exceed $6 billion, making it the most expensive storm to date in Canada.

This hydro pylon near Constant, Quebec was downed in the 1998 ice storm.

Saguenay Flood

THE SAGUENAY–LAC-SAINT-JEAN region of Quebec is spectacularly beautiful. Lac-Saint-Jean is fed by dozens of rivers that drain the area, and it in turn empties into the Saguenay River, which flows through a marvellous canyon to the St. Lawrence. Nearly 300,000 people live here, in a region that is not only the centre of agriculture and forestry but also the heartland of Quebecois culture.

In 1996, a weather phenomenon known as an omega block was plaguing eastern Canada. An omega block occurs when strong low pressure anchors itself over a region and cannot migrate eastward because a high-pressure system is blocking its path—in this case, a Bermuda high over the western Atlantic. (The event is called an omega block because the jet stream follows a route around the high and low that takes on the shape of the Greek letter.) Until that high weakens, the low stays put and produces rain.

Indeed, the spring and summer of 1996 were among the wettest ever in many parts of eastern Canada: April, May and June saw at least twice the average rainfall, and reservoirs in the Saguenay area were filled nearly to overflowing. It rained almost every day from Saint-Jean-Baptiste Day (June 24) until mid-July, and by July 19, the accumulation for the month stood at 138 millimetres, already 31 millimetres over the monthly average. That Friday, the

skies dumped another 62 millimetres of rain, followed by 56 more on Saturday. The two-day rainfall supplied as much water as goes over Niagara Falls in four weeks.

The effect of all this rain was magnified by the topography. The Saguenay region is known as a graben, from the German word for "trench." A graben is an area of sunken land between two faults in the Earth's crust. As the rain teemed down, the water was funnelled to the lowest part of the valley, where the Saguenay and Chicoutimi rivers meet. In less than 48 hours, a month's

Water stays in the atmosphere for an average of 11 days before falling as precipitation.

worth of rain flowed into the already swollen streams and tributaries of these rivers. More than 2,000 dams, barriers and structures control water flow in the region, but most had been built before strict engineering codes were imposed. It wasn't long before rivers and streams began overflowing their banks and water spilled over the tops of the dams and dykes.

On Saturday, July 20, workers opened spillways wider to allow more water to exit from the rapidly rising Lake Kenogami. As the water was drained from the lake, engineers knew it would cause flooding downstream, but they had no choice. It didn't take long for the earthen dam at Lac Ha! Ha! to rupture, tearing a 20-metre trench through the forest as it was pushed forward by 30 million cubic metres of water.

The rising crest of water consumed trees, power lines, bridges and vast stretches of road. The village of Boilleau was swamped with water and uprooted trees. The torrent then continued to the town of La Baie, where it exploded into buildings with the force of a massive bomb, sweeping people away with their homes. Fortunately, help came quickly from CFB Bagotville, just down the road. Ten people lost their lives, but 16,000 were safely evacuated. When the water was finally gone, 900 cottages and 1,718 houses were destroyed or severely damaged. In the four weeks that followed the deluge—which was Canada's first billion-dollar weather disaster—the military served 40,000 meals to people who no longer had homes to which they could return.

During the July 1996 Saguenay flood, the Chicoutimi River burst its banks and tore through the town, forcing 12,000 to flee their homes.

Edmonton Tornado

THE LAST WEEK of July 1987 was hot and muggy in Edmonton: the humidex had it feeling like the high 30s in the afternoons. It was vacation time for many, and the provincial capital was just coming off a successful Klondike Days celebration. On Friday, July 31, it really did feel like the dog days of summer.

The humidity was as result of a deepening area of low pressure over Alberta. This surface-level low was drawing increasingly moist and warmer air into Alberta and Saskatchewan. A ridge of high pressure aloft in the atmosphere acted as a lid, keeping the heat near the surface. By mid- to late afternoon, thunderstorms would begin to grow west of Edmonton and then slowly drift eastward, their potency kept in check by the upper ridge. It is like this every summer in Alberta. The climate east of the Rocky Mountains is highly conducive to powerful storms: in

an average summer, Alberta will witness the development of 10 tornadoes and 52 days with hail-producing thunderstorms.

The previous day had seen thunderstorms develop in mid-afternoon, bringing heavy rain and winds over 80 km/h. By late Friday morning, it was still cloudy and the winds had become more southerly. At 2:40 in the afternoon, severe thunderstorm warnings were issued by Environment Canada. Most Albertans are accustomed to these events, but it soon became clear that this was not a typical summer thunderstorm. Just before 3 p.m., a tornado was sighted near Leduc, 25 kilometres south of Edmonton, not far from the international airport. Tornado warnings were immediately broadcast to the surrounding area.

Between Leduc and Edmonton, the tornado grew to F3 (winds between 254 and 331 km/h) as it brought down power lines and damaged buildings. The community of Mill Woods was the first large urban area to see significant damage as the twister plowed through the community's winding roads and cul de sacs, built in the early 1970s. The tornado roared north, staying east of the Calgary Trail and west of Sherwood Park. It continued to grow in strength, and by the time it hit Refinery Row, it was rated as an F4, with winds between 332 and 418 km/h. Winds of that speed will toss vehicles like tin cans and leave very few buildings standing (only two percent of all tornadoes ever grow this strong).

Scenes of devastation after the 1987 Edmonton tornado; the mobile home in the background was flipped completely upside down.

Tornadoes can turn the sky black with dirt and debris as they tear through the landscape.

The tornado was now 30 minutes old, and it had already taken 12 lives.

As the Edmonton tornado continued to travel north, it began to lose some of its energy. By the time it struck the Evergreen Trailer Park at 4:25 p.m., it was estimated to be at F2. Even with this decrease in strength, the winds were still between 181 and 253 km/h. Another 15 people would die in the trailer park.

The tornado left a swath 37 kilometres long and up to a kilometre wide along the east side of Edmonton. In addition to the 27 killed, 600 people suffered injuries, and 1,700 were rendered homeless.

On May 31, 1985, a series of tornadoes touched down in Barrie, Ontario, killing eight and injuring 160. The 200-km path left behind was the longest recorded in Canadian history.

Prairie Drought

PRAIRIE FARMERS RELY on sufficient winter snow and adequate spring and summer rain for their planting to be successful. The margin between a good year and a bad year is measured by the weather, and to live through the bad years is to endure hardship on a large scale. Unlike most other weather disasters, a drought is hard to pin down to one event: it is a series of climactic conditions that prevail over an extended time. Yet drought can be one of the most costly and widely felt weather disasters: when crops fail and livestock can't be fed, then drought will touch all of us.

The drought on the southern Prairies that eventually peaked in 1988 began with decreased rain and snowfall that persisted for several years. By the time January 1988 arrived, there was a growing concern among the agricultural and scientific communities that a widespread drought was underway in central North America.

Let's look at Saskatoon and Estevan, part of Saskatchewan's grain belt. On average, Saskatoon receives 350 millimetres of precipitation each year, while Estevan receives a combined 433 millimetres. In 1984, there was a 10 percent drop in precipitation, enough to cause some minor stunting of crops and slightly

shorter yields. The weather grew more worrisome in 1985: snow was down about 25 percent in central Saskatchewan, meaning the soil received less moisture from melting in the spring. Rain and snow returned to normal in 1986, but the following year would see a disaster unfold in slow motion. Winter snow was 40 percent below average, the rain was a third or more below normal, and the temperature throughout most of the Great

The hottest Canadian temperature, 45°C, was recorded at both Midale and Yellowgrass, Saskatchewan, on July 5, 1937.

Parched Prairie farmland during a period of drought.

Plains was unusually warm.

The problem with decreasing moisture in soil is that the soil bakes like clay in a kiln. The rich soil blows away on the western wind, and the earth becomes increasingly difficult to till. When heavy rain does fall on this hard and dry soil, it runs off before it can be absorbed by the crops.

As the summer of 1988 arrived, the Prairies were in the midst of a full-blown drought. Rain and snow were down over 20 percent, and temperatures in Saskatoon were climbing into the mid-30s in May. By June, the city set an all-time record for heat: 40.6°C. Reservoirs fell to critical levels, and water conservation became necessary. The year was one of the worst in a quarter-century for forest fires in all three Prairie Provinces. In the central United States, 1988 was hotter and drier than it had been during the Dust Bowl of the 1930s.

The financial toll was estimated to be $3 billion dollars, making the 1988 drought the most expensive weather disaster in Canadian history. Farmers lost tens of thousands of hectares of grain, beans, corn and animal feed. Malnourished livestock was sold far below cost, while many animals simply starved or died of thirst. At least 10 percent of all farms went bankrupt, and 10 percent of all workers in agriculture left the business.

Hurricane Juan

Residents of Halifax, Nova Scotia, survey the damage left by Hurricane Juan on September 29, 2003.

MARITIMERS ARE FAMILIAR with tropical storms and hurricanes—they experience them every year—but the one that arrived just after midnight on September 29, 2003, would be different. In Nova Scotia, residents were about to see things they'd only heard about in sea stories.

Hurricane Juan began as a tropical depression on September 24 about 300 kilometres southeast of Bermuda. A day later, it was tracking north and grew strong enough to be classified as a tropical storm. By September 26, Juan was a Category 1 hurricane with sustained winds over 119 kilometres per hour. It would remain a full-fledged hurricane as it transited Nova Scotia, the Northumberland Strait and Prince Edward Island. To make a journey of nearly 225 kilometres inland while maintaining winds of hurricane strength is a feat that few storms manage this far north.

Forecasters were almost certain that Juan would make landfall somewhere in Nova Scotia before the weekend was out. I recall Shelly Steeves, The Weather Network's reporter in Atlantic Canada, sending reports from Halifax as people made preparations. By Sunday evening, the outer bands of Juan began to lash at the Atlantic coast. An oil rig in Halifax Harbour recorded a wind speed of over 180 km/h: that's how hard the gale was blowing against the high rises of the city and on the bridges that span its harbour. Winds that powerful create a storm surge: a welling of seawater caused by the force exerted on the ocean by a constant, powerful wind.

The storm surge began along the coast at 9:45 p.m. Offshore marine buoys measured rises of nearly 20 metres, while a buoy at the mouth of Halifax Harbour measured a nine-metre rise in the sea level. High tide arrived around 10 that evening, and water continued rising until one in the morning. On the waterfront of both Halifax and Dartmouth, there was widespread flooding as the wind literally blew the ocean onto the coast.

At picturesque Peggys Cove, a wall of

A resident of Bedford, Nova Scotia, finds a sailboat has been thrown ashore in her front yard during Hurricane Juan.

water nearly two metres high swept into the community. Peggys Cove faces the sea to the west, and this wave cascaded in from the east. That means the wave had to rise at least seven metres from the sea, cross over hills and travel nearly half a kilometre across a peninsula to arrive at Peggys Cove. The people who witnessed the surge said the water rose this way as many as 12 times during a 30-minute period. They had seen similar occurrences in the past but never to this magnitude.

The eye of Juan was 35 kilometres wide as it crossed the Nova Scotia coast. The northeast quadrant of a hurricane bears the strongest winds, and this most destructive edge struck greater Halifax. The greatest damage extended from Prospect east to Clam Harbour. The winds downed trees and power lines throughout the province. By Monday morning, hundreds of thousands of Maritimers were without power, and many would wait two weeks for electricity to return.

As Juan crossed Nova Scotia, it was weakening, but it was still a hurricane. By 3 a.m., it was at the Northumberland Strait, passing directly over the Confederation Bridge. The storm surge flooded the waterfront in Charlottetown, and the airport reported sustained winds of 95 km/h with gusts to nearly 140 km/h.

Wind speed in a weather report is an average velocity recorded during a two-minute period.

Calgary Hailstorms

On July 14, 1953, a hailstorm pounded Alberta, killing thousands of birds. Several days later, it happened again.

THE AREA FROM Lethbridge north through Calgary to Edmonton is known as Hailstorm Alley, and it's prone to some of the world's most frequent and dangerous hailstorms. In the summer in Alberta, the genesis of low pressure and the region's topography combine to provide ideal conditions for potent thunderstorms. These can produce hail when water journeys up and down through a thundercloud, freezing and growing in size with each cycle.

On July 16, 1996, clusters of thunderstorms grew and pounded the city. These storms grew out of an elongated cold front that was sagging southward, the front extending from Alberta to Manitoba. Shortly after 6 p.m., the skies opened up and torrents of rain and hail fell on Calgary—and the storm lasted for several hours. Hailstones the size of grapefruits clogged storm drains. The floodwater rose up to knee level, and many roads were closed because of the high water. Power and telephone poles came down and emergency 911 services were temporarily knocked out. The hail damaged windows, buildings and thousands of vehicles.

Farther east, Winnipeg was also pounded with hailstones as large as tennis balls, as well as heavy rains. There was extensive damage to crops, property and vehicles, and minor flooding was recorded as storm drains became blocked with hailstones. The combined cost of that week's hailstorms would top $300 million.

Just over a week later, it happened again. On July 24, thunderstorms arrived in Calgary at four in the afternoon, bringing hailstones as big as oranges. The damage to homes and vehicles was just as extensive as it had been only eight days earlier.

An impending hailstorm near Erickson, Manitoba.

6

Climate Change

OUR CLIMATE HAS been in a constant state of change since Earth cooled enough to support a liveable atmosphere some 3.5 billion years ago. These variations in our planet's climate occur because of several factors, and they lead to changes in local weather patterns and ecosystems. The changes can be spread over enormous periods of time. Think about plate tectonics: as the continents move over the millennia, the physical geography of Earth is altered. When mountains and oceans evolve, climate and weather patterns change as well, through ice ages to interglacial (warmer) periods and back again. It is thought that these cycles are driven by the changing angle of the Earth's axis and the path of its orbit around the sun. Today, our planet's axis is tipped 23.5 degrees and the North Pole points almost directly at Polaris, the North Star. But in cycles that take tens of thousands of years, the angle of the axis varies between 22 and 24.5 degrees, and Earth wobbles like a spinning top. In an even slower cycle measured in hundreds of thousands of years, the orbit of our planet changes slightly, becoming more and less elliptical. These oscillations radically affect the amount of solar radiation reaching the surface.

The icebreaker *Kaptain Khlebnikov* sails near Baffin Island through rapidly melting ice.

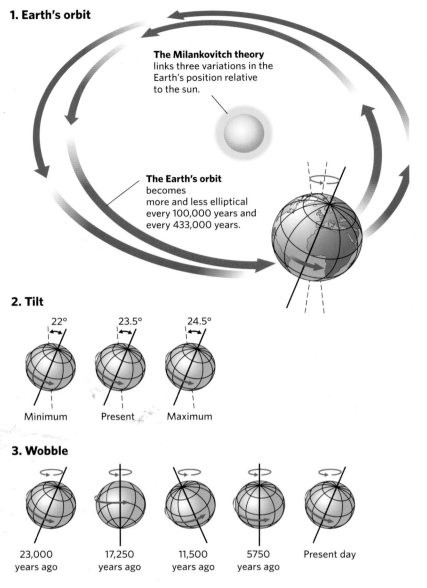

1. Earth's orbit

The Milankovitch theory links three variations in the Earth's position relative to the sun.

The Earth's orbit becomes more and less elliptical every 100,000 years and every 433,000 years.

2. Tilt

22°   23.5°   24.5°

Minimum   Present   Maximum

3. Wobble

23,000 years ago | 17,250 years ago | 11,500 years ago | 5750 years ago | Present day

If the Antarctic ice cap melted, it would cause global sea levels to rise by 65 m.

The last ice age was at its peak about 18,000 years ago. During that period, ice sheets as thick as three kilometres covered what is now Canada and extended into the northern third of the United States, as well as Scandinavia, the northern British Isles and much of what is now northern Russia. South of the equator, Argentina, southern Australia and New Zealand all lay under ice. The oceans adjacent to these areas froze over, and sea levels around the world fell. As

the cycle continued, a warming ensued, and by 7,000 years ago, much of the ice had retreated, and the oceans were again rising with the meltwater.

We are only now beginning to measure other cycles that affect our climate. El Niño and La Niña, the temperature fluctuations in the Pacific Ocean, have only become understood in recent times, and more patterns will emerge as we further study the climate of our planet.

While some cycles occur over periods ranging from a decade to several millennia, there are also instances of sudden change to our climate. A comet or asteroid impact can change the entire planet's climate in just a few years—a mere instant in geologic time. The most famous such event occurred about 65 million years ago, causing the extinctions of countless species, including the dinosaurs. When a large object strikes Earth, the collision throws a massive cloud of dust and debris into the atmosphere, blocking sunlight from reaching the surface and quickly changing climate patterns. Large volcanic eruptions can do the same thing: after the eruption of Krakatoa in Indonesia in 1883, global temperatures dropped 1.2 degrees (C) for five years.

We find evidence of this changing climate in fossils, ice cores and sediments. As we study this ancient material, we have discovered climatic cycles and patterns that illustrate how our planet evolves. More evidence emerges when we look back at human history. The first evidence of agriculture occurs about 10,000 years ago, an era that coincides with the end of the last ice age. Archaeologists have found ancient cave art in the Sahara

Desert that illustrates cattle being raised in the area dating back over 4,000 years, evidence that the climate there was once lush enough to support an agricultural society. The decline of Mayan civilization a thousand years ago has been linked to an extended drought brought on by a sudden climate change. At the same time, thousands of kilometres away, Norse settlements appeared along the coast of

In 1883, the thick clouds of gas and ash that spewed from Krakatoa deposited high levels of sulfur dioxide into the stratosphere, saturating high-level cirrus clouds and causing them to deflect the rays of the sun. This resulted in a cooling of the entire planet, causing chaos in weather patterns until 1888, five years after the disastrous eruption.

Scientists have discovered that, as a worldwide average, Wednesday is the warmest day of the week.

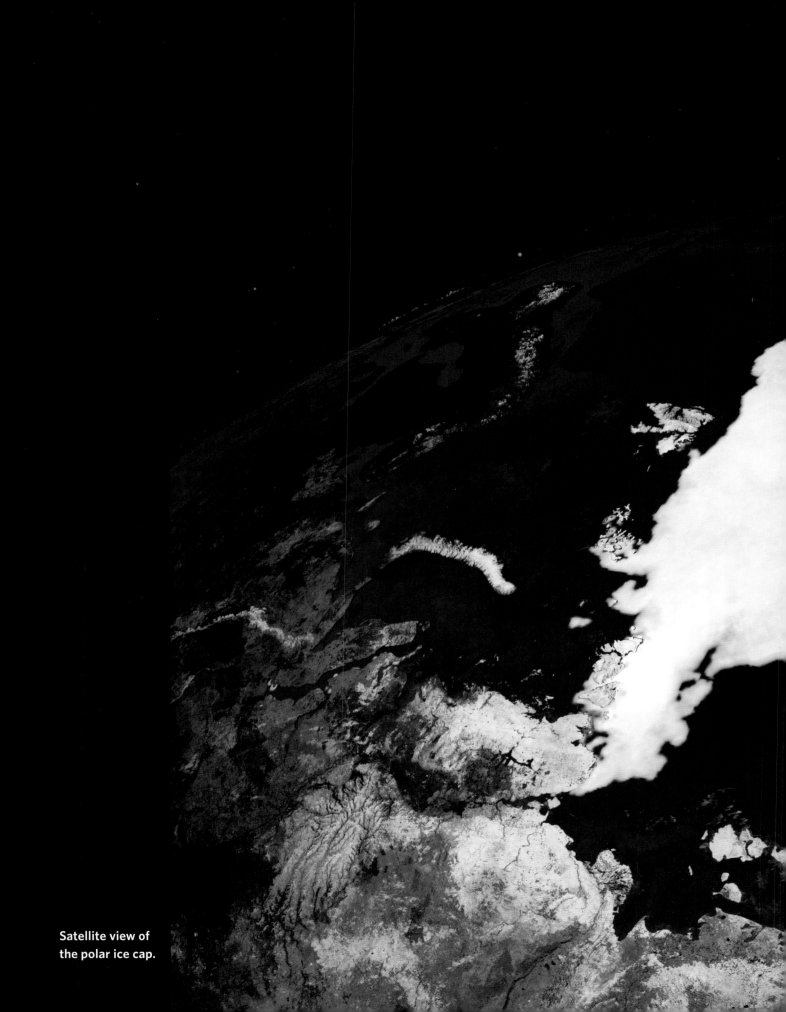

Satellite view of
the polar ice cap.

Greenland and Labrador, areas that are today inhospitable.

In other words, climate change has always occurred naturally, and human beings have always found ways to adapt. The concern today, however, is that Earth is warming at a much more rapid rate than we have ever seen before. The changes are not unfolding over thousands of years; they are occurring over the span of a human lifetime, and the phenomenon may have ramifications for all life on Earth.

The last decades of the 20th century saw the average global temperature rise at rates that we have never before witnessed since weather recordkeeping began, and the average temperature on our planet continues to increase. There is no debate that we are witnessing rapid global warming, and a world body of scientists assembled by the United Nations agrees that much of this warming is directly attributable to human activity. Our actions and activities—such as industry and agriculture—are accelerating a natural cycle.

The atmosphere is made up of many gases, including oxygen, carbon dioxide, ozone and water vapour. These gases absorb infrared radiation from the sun and radiate it to the surface, warming our planet. This system was fairly stable until about 200 years ago, when the industrial age was born and humans began to burn coal and other fossil fuels in great quantities. By the 20th century, as technology and medicine advanced our societies, the human population began to grow at an astonishing clip. There were about a billion people on the planet 200 years ago; today, there are more than 6.6 billion. That huge population requires a lot of food and energy.

Through most of history, the human footprint on Earth has been minimal. We grew enough food to support ourselves and burned enough wood to stay warm. None of these activities were enough to bring about large-scale changes in climate patterns. In the past two centuries, however, our large-scale use of fossil fuels has dramatically increased the amount of carbon dioxide in the atmosphere. At the same time, deforestation to create more agricultural land is removing the trees required to absorb that carbon dioxide. More CO_2 (as well as other so-called

Melting Ice Caps

In the past 30 years, more than 20 percent of the polar ice cap has melted away, and scientists speculate that, without a change in emissions levels, it will be completely gone within 100 years.

Global Temperature (Land-Ocean Index)

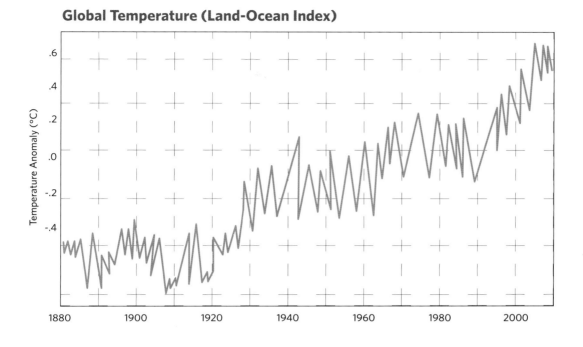

greenhouse gases, such as methane and nitrous oxide) mean more of the sun's heat is trapped at the surface. This is slowly increasing the temperature around the world, especially near the poles.

Scientists studying climate change agree that this warming pattern will continue: they project that the average global temperature will rise another four degrees by 2060, and by as much as nine degrees in both the Arctic and Antarctic regions during the same period. We are already witnessing the effects in the Canadian North: the ice cover on the Arctic Ocean is retreating farther north every summer, and it is projected that there could be an ice-free summer within the next 50 years.

The removal of that much ice will change weather patterns in the entire northern hemisphere for a number reasons. Ice reflects about 80 percent of the energy it receives from the sun; the dark sea, by contrast, reflects only about 10 percent of that heat and absorbs the

rest. Warmer seas lead to even more melting ice, creating a feedback loop that results in ever-warming temperatures. The addition of massive quantities of fresh water into the oceans will also alter salinity levels, further changing weather patterns. The melting of ice caps and glaciers will also increase the volume of water in the ocean, and coastal lowlands will be prone to flooding. Finally, the thawing of the decomposed plant matter in arctic bogs releases massive amounts of carbon dioxide into the atmosphere.

Warmer oceans and warmer air masses generate weather that is more volatile and difficult to predict. As the atmosphere churns with warmer air and increased moisture, hurricanes and other storms will become more frequent and more powerful. The ability to predict how the weather will behave is based on historical data: when we marry that experience with the information collected today, we can make an accurate projection as to what

On a clear, still summer evening, urban temperatures can be several degrees warmer than in the suburbs.

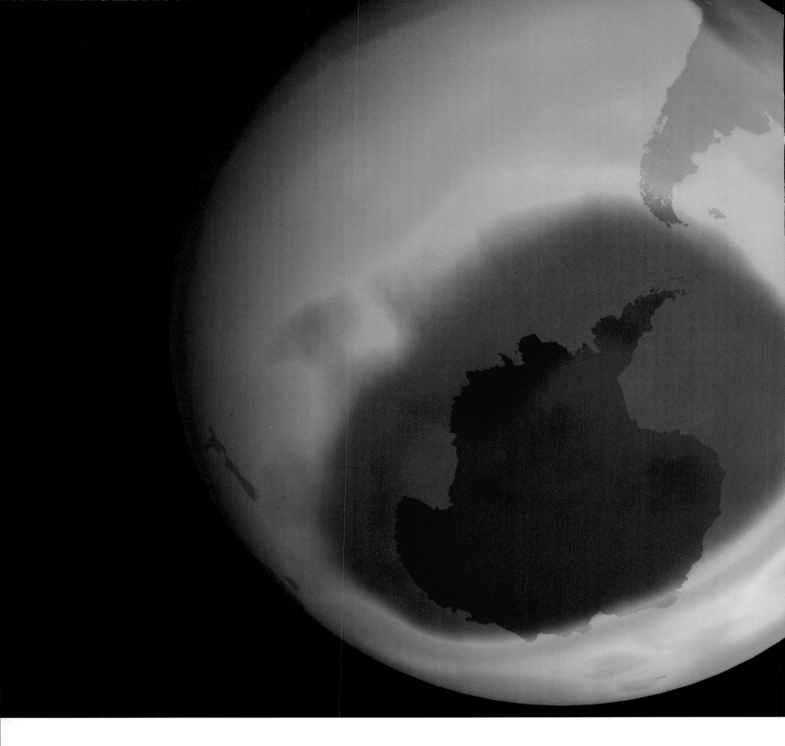

This image of the hole in the ozone layer over the Antarctic taken in 2002 shows it at its largest on record.

will happen next. But the rapid changes in climate and weather patterns today make next week's forecast much less clear.

The only way to slow the warming is to reduce the amount of carbon dioxide in the atmosphere. As a species, we have the intelligence and ability to do this. Some of the damage is already done, but

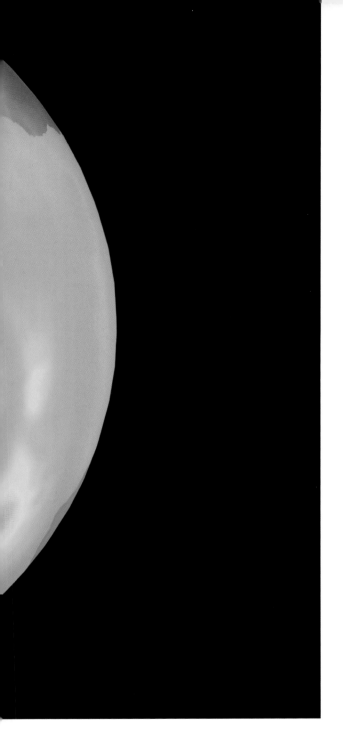

children. The United Nations Inter-governmental Panel on Climate Change has suggested ways that we can lessen our impact on the planet. Doing so involves dramatically reducing our dependence on carbon-based fuels and radically increasing initiatives to reforest Earth.

There are also many simple things we can do on a personal level. If every lightbulb were changed from standard incandescent to energy-efficient fluorescent, we would reduce global emissions by nearly 10 percent. We can also strive to become carbon neutral—that is, for every tonne of CO_2 we emit, we should do something to take back a tonne. By reducing our consumption of meat by 25 to 50 percent, we would drastically cut greenhouse gas emissions. Ending our reliance on oil and other fossil fuels and using wind, solar, hydrogen, geothermal and tidal energy would have the largest impact in the amount of carbon dioxide we put into the atmosphere.

The question, of course, is whether we have the will to act. History offers at least one encouraging precedent. In 1985, the media around the world reported that there was a large hole in the ozone layer. Ozone (O_3) is a gas that occurs in the stratosphere, where it absorbs ultraviolet radiation and protects us from the cancers that high doses of this radiation cause. In the 1970s, scientists began to understand that ozone reacts with some of the molecules that humans release in the atmosphere, notably chloro-fluorocarbons (CFCs), which were once widely used in fire extinguishers and as refrigerants. Once the CFC molecules reach the stratosphere, they are broken

slowing the pace of climate change can buy humankind time to adapt to future changes. By reducing carbon dioxide emissions now, we will ease the warming trend and evolve toward a cleaner and more efficient society. By tending to our planet's forests and natural gardens we will create a healthier home for our

apart by ultraviolet radiation. The newly liberated chlorine atoms then react with and destroy ozone molecules, allowing UV radiation to pierce the atmosphere. By the mid-1980s, the ozone layer over the South Pole was so thin that a "hole" developed and grew to the size of the continent of Antarctica. Soon, there was a noticeable increase in UV radiation at the surface.

The world reacted swiftly to this environmental crisis: in 1987, countries around the world created the Montreal Protocol and agreed to reduce or eliminate the manufacture of products that release CFCs into the atmosphere. As a result, global production of CFCs has dropped 40 percent and the hole over the South Pole appears to have stopped growing. It will take many decades for nature to scrub the CFCs from the atmosphere, and we don't know for sure whether the ozone hole will seal itself. But the Montreal Protocol is an illustration of our ability to collectively agree to repair some of the damage we have done to our planet. Only human ingenuity and tenacity will help us weather this latest storm.

Q Which Canadian community has experienced the greatest temperature change?

A A chinook blew into Pincher Creek, Alberta, on January 27, 1962. The temperature went from –19°C to 22°C in one hour, a rise of over 40 degrees.

Receding ice cap
in Greenland.

Index

Photo Credits